THE BOOK OF

THE MACHINE GUN

Photograph by Gale and Polden, Ltd., Aldershot.

Machine-Gun Section, the Royal Munster Fusiliers, at Aldershot (of the Original Expeditionary Force),

THE BOOK OF THE MACHINE GUN

BY

MAJOR F. V. LONGSTAFF

LATE 5TH BATTALION (TERRITORIAL) THE EAST SURREY REGT.

AND

A. HILLIARD ATTERIDGE

LATE CAPTAIN, LONDON IRISH RIFLES

LONDON

HUGH REES, LTD.

5 REGENT STREET, S.W.

1917

AUTHORS' PREFACE

THE following pages represent the result of a year and a half of work. The book was projected by Major Longstaff, and taken in hand after he had left the Service, through ill-health, in the spring of 1915. He then began the collection of material and the compilation of the bibliography. In the following July, Captain Hilliard Atteridge became his joint author in the work. The authors could not have completed it without the co-operation and help which they have received from a large number of sources, and which they must here gratefully acknowledge.

Major Wylly, the Librarian of the Royal United Service Institution, has given great assistance in finding and tracing papers, books and pamphlets of all kinds, and in searching the files of the Journal for reports of lectures. The authors have also to thank the Council of the R.U.S.I. for permission to reproduce extracts from the Journal, and but for the Library of the R.U.S.I., the work could not possibly have been produced in the time.

The work of Commandant Lavau, now Director of the French Machine Gun School of Instruction at Vincennes, has been of the greatest help in collecting the opinions and tactical ideas of continental writers, and the authors have to thank the Commandant for permission to reproduce important passages from his work.

The authors have to acknowledge the assistance received from the Secretary of the Royal Artillery Institution, the Secretary of the U.S. Army General Staff, and the Librarian of the Royal Geographical Society, Mr. E. Heawood, M.A. Major H. W. Todhunter, of the School of Musketry, has supplied some useful historical information. Major T. C. Bamfield, the Commandant of the Canadian Machine Gun School, has given useful assistance and advice, and Captain H. F. Meurling, formerly Instructor in Tactics at the Canadian Machine Gun School, has given important help in working out the question of fire tactics.

Major F. B. McRae, 12th Battalion C.E.F., and Lieutenant C. A. Bradbrooke (formerly Adjutant of the Canadian Armoured Car M.G. Brigade of Ottawa), both of the Canadian Machine Gun School, and Lieutenant-Colonel C. R. Crosse, M.V.O., and Lieutenant-Colonel J. Barlow, M.V.O., of the National Rifle Association, have also given valuable assistance.

Assistance has also been received from the Directors of Messrs. Vickers in the way of information of various kinds, letters, books and photographs. Messrs. Kynochs have kindly supplied information, the Directors of the French Hotchkiss Company have also kindly supplied books and information in letters, the same help has been received from Colt's Firearms Manufacturing Company, and Colonel Lewis (late U.S. Artillery) has given much information as to his gun.

The authors have to acknowledge help in various ways from the Editors of the following journals: the *United Service Gazette*, the *Navy and Army Gazette*, the *Engineer*, *Engineering*, the *Scientific American*, the *Daily Graphic*, the *Illustrated London News*, and the *Sphere*.

Mr. R. K. Murray, of Plymouth, is thanked for naval and military historical information, and Company-Sergeant-Major Bruce White, Artists' Rifles, for historical notes on aids to beginners in the mechanism of the Maxim.

Help in the difficult work of tracing and obtaining rare and out-of-print books and pamphlets has been given by Mr. Hugh Rees; Messrs. W. Clowes and Sons; Sifton Praed and Co., Ltd.; Foster Groom and Co.; Hachette and Co.;

Mr. A. F. Bird and Mr. Francis Edwards, of London; Mr. L. W. Jones, of Folkestone; Mr. W. H. Long, of Portsmouth; Mr. H. G. Long, of Southsea; Mr. J. S. Amoore, of Wimbledon; Messrs. Gale and Polden, of Aldershot; Mr. S. H. Tremayne and Mr. W. J. Butland, of Plymouth.

The following have to be thanked for assistance in finding and providing photographs for illustrations: Messrs. Gregory and Seeley, Ltd.; Messrs. Underwood and Underwood, Ltd.; the Record Press; and the Sport and General Agency, all of London. Messrs. Gale and Polden, of Aldershot, and Messrs. H. Symonds and Co., of Portsmouth; Messrs. Gibbs, of Pirbright. Mr. J. H. Steward, of London, has kindly furnished an illustration of his clinometer.

The eighty-five illustrations have been selected from a collection of over two hundred. The aim has been to make each picture tell its own story with the help of the title and the notes at its foot. The series has been arranged in chronological order, so as to give the student a graphic presentment of the progress of machine-gun matériel. Where it has not been possible to ascertain the precise date of a photograph, the year of issue of the machine gun or mounting it shows has been taken as a guide. The latest information as to many of the guns will be found in the explanatory notes to these illustrations, which include supplementary details received while the book was in the press. The only abbreviation that needs explanation is "R.C.A.M.Gun" for "Rifle Calibre Automatic Machine Gun."

F. V. L.
A. H. A.

LONDON,
November, 1916.

CONTENTS

APPENDICES

LIST OF ILLUSTRATIONS

MACHINE GUNS

CHAPTER I

INTRODUCTORY

THE soldier is by nature conservative. Mature age and high command usually go together, and it is the exceptional man who, as years increase, maintains the openness and elasticity of mind that welcome new ideas. Among the younger men, until very recent times, an officer hardly improved his prospects by being a seeker after novelties. It was a sounder policy to accept the existing regulations and the traditional methods as sufficiently near to perfection for all practical purposes.

One can trace the influence of this cautious conservatism far back in the development of armaments and tactics. Long after the invention of gunpowder the knightly soldiers of the time regarded gunnery as a mechanical art to be left in the hands of mere tradesmen. They did not trouble themselves to learn anything about it, and were content to hire the professional artillerists to assist them in battering down castle and city walls, or to fire a few shots at the beginning of an engagement. The result was that it required more than three centuries of slow evolution to convert the clumsy bombard of the middle ages into the field gun of the first part of the eighteenth century. It was nearly another hundred years before the guns were worked in permanently organized batteries.

The musketeer had for at least two hundred years to depend upon the pikemen to protect him. The pike remained an infantry weapon long after a Frenchman struck on the fairly obvious plan of fixing a steel blade on a musket to enable the musketeer to defend himself at close quarters.

Sportsmen were using percussion caps for long years before any army gave up the old flintlock, and long after the percussion cap had obtained reputation inventors were vainly trying to persuade the Governments that a capped cartridge was not too dangerous for military use—this, too, at a time when capped central fire cartridges were being freely used in sporting guns. Every War Office in Europe, with only one exception, made a long fight against the introduction of the breechloader. The Prussian needle gun had been used with marked effect in three wars before it dawned upon the other European Governments that the breechloading rifle was better than the muzzleloader.

Rifled field guns were first used by artillery on the French side in the war of 1859. In the following year some of our field artillery was armed with the new rifled guns. They were not at all popular among the older officers. Sir Alfred Turner, then a young subaltern in the Dover garrison, tells how, after the annual inspection of 1860, he heard a conversation on the subject between the Inspecting General and the local Commanding Officer of the artillery, both of them men who had begun their service in the Army under Wellington. In reply to the question from the Inspecting Officer, "What do you think of the new guns?" the other replied, "I can't say I think much of them. We won Waterloo without them, and what do we want with them after that?" Perhaps this was an exaggerated case of military conservatism.

But the same conservatism worked in another direction to produce very unfortunate results, even after a new weapon had been officially adopted. It insisted on using the improved weapon with the same tactical methods which had been employed with the weapon it had displaced. Thus, for instance, on the Crimean battlefields our infantry was armed with the long-ranging Enfield rifle, but it was brought into action against the Russians at the short range of the old Brown Bess. The effect on the enemy, who moved in massive columns, was deadly enough, but at the same time we incurred heavy loss from the Russian musketry fire—a loss which might have been considerably reduced if the officers of the day had realized the advantage the new weapon gave them, and had had enough imagination to prepare for the attack at close quarters by heavy rifle fire at medium ranges.

The enormous losses incurred by the Germans in the first stage of the war of 1870—for instance, those of the two brigades in the attack at Vionville, and of the Prussian Guard Corps at St. Privat—were the result of using against the breechloader the tactics which were good enough against the old musket. On the battlefields of 1866, and even of 1870, rifled field guns of long range were brought into action side by side with the infantry.

The tradition of the old times, before the fire of infantry had become really effective, and when armies were largely made up of mounted troops who won battles by riding down and breaking up the opposing infantry, was a tradition that lingered on almost to our own times, and led to hundreds of disastrous cavalry charges, in which brave men threw away their lives in trying to achieve the impossible.

In our days, thanks probably to the rapid changes produced by scientific discovery and invention in the ways of civil life, the innovator has had a better chance both in armies and navies. Military conservatism, though still a power, is no longer all-powerful. We can trace its influence, however, in the record of the introduction of the machine gun into most armies. First it opposed the innovation; then it retarded the development of the new weapon by insisting on using it with tactical methods that were in complete discord with the spirit of machine-gun warfare.

It was first used as if it were a new and very inferior kind of cannon. Having failed to do what it was never meant to accomplish, it was regarded as a failure, and then taken up again as a machine that might be of use to stop the rush of ignorant savages, but could have no place in regular warfare between civilized nations. When, thanks to a few enthusiasts, it began to find its way into European armies, it was officially decided that it was merely a defensive weapon, that might be used with advantage in special circumstances. Then came the gradual realization of the fact that it might find a place in the attack, and it was given in small numbers to cavalry and infantry as a kind of auxiliary weapon to be used more or less like the light battalion guns that were dragged into action in the intervals of the fighting line in the days of Gustavus Adolphus and Frederick the Great before artillery had won a place of its own on the battlefield. In our own Army, in this stage of the evolution of the new weapon from which we are now beginning to emerge, the handling of the guns and the training of their detachments was in most cases the temporary duty of a young officer, who made it either a routine occupation, or became interested in it and worked at it as a hobby. In colonial and frontier wars, the importance of the gun was recognized, but in the years before the Great War, the unwritten rule at manœuvres appears to have been to "stick them in somewhere." Commanding Officers seldom gave them any very definite place in their scheme, and umpires paid little attention to them. The machine gun had not come into its own. It was an accidental appendage to a fighting unit, and chance decided whether any intelligent use was made of it.

But there are now signs that a change is coming for the machine gun as great as that which won its recognized place for artillery in the wars of the French Revolution and the First Empire, when Napoleon won battles by smashing his enemy's front with massed artillery at short range to prepare for the decisive advance of his infantry, or at Wagram used a line of eighty guns to close a gap in his array and hold it by their unsupported fire. Artillery at last became an arm at least as important as the infantry or the cavalry, and was recognized as a battle winner.

We used to speak of the "three arms"—infantry, cavalry, and artillery. The growing importance of military engineering raised the claim of the engineers to be recognized as a fourth arm, and the coming of the aeroplane and the airship gave us a fifth. Perhaps we shall soon recognize that there is a sixth arm. The French call machine guns *mitrailleuses*, and a French specialist has suggested that, as we speak of artillery, we should also speak of *mitraillerie*, but the name does not matter much so long as we have the thing.

If we are to make the fullest use of the new arm we must understand not the mere mechanism of the machine gun, but, if we may use the word, its spirit. We must realize what are the powers of the gun and the conditions under which they can best be brought into action. The mechanical engineers have given us in various forms the means of enabling two or three men to deliver a fire equivalent to that of fifty or sixty rifles, with deadly effect. The use of such an enormous power as this should not be the mere temporary occupation to which a few men and officers are assigned for a while. Rather it should be the life work of a machine gunner, as the handling of his guns is that of the artillerist, and its supervision and development should be the mission of a special section of the Imperial General Staff. Only thus shall we have a body of officers and men inspired with the common tradition, and capable of producing the highest results.

We have to build up a working theory of the machine gun. As a basis for it, we have the experience of some sixty years, during which the gun has been evolved, and an immense mass of records of facts and attempts to produce practical lessons from them; but to the average man much of this is inaccessible. Some of it is buried in the files of professional journals, or embodied in books and memoirs that are out of print or have become rare. Some of the most valuable matter is to be found in foreign publications, and the number of those who can read foreign languages easily is not large. It will be a useful work, therefore, to endeavour to bring together within moderate limits of space all that is most useful in this literature of the machine gun. As a basis for the practical consideration of questions of matériel, tactics, organization, and training, it will be well first of all to tell the story of the evolution of the new weapon, and to bring together some record of its actual use in war.

CHAPTER II

THE EVOLUTION OF THE MACHINE GUN

A MACHINE GUN may be broadly described as a small calibre firearm, with one or more barrels fitted with a mechanical contrivance to secure a rapid fire. This is done in two ways: (1) By the movement of a handle which actuates the machinery and continuously reloads and fires the weapons, or (2) by an automatic device which uses either the recoil of the gun or part of the explosive force of the charge to actuate machinery, which reloads and fires a succession of cartridges once the operation has been started by firing the last shot.

A further point must be noted in order to exclude repeating rifles and pistols from this rough definition. The machine gun is a weapon of such power that it must be fired from a stand of medium weight for rapid and accurate laying and to ensure effective fire.

The machine guns of the present day are the result of a series of inventions beginning a little more than fifty years ago. But from the earliest days of firearms attempts were made to produce weapons of this kind. The oldest types of machine guns, known as *ribaudequins* or *orgues*—that is, organ guns—were nothing more than groups of six or ten musket barrels mounted side by side on a frame, and either fired by each barrel being provided with its lock, or by a single lock firing a quick-match, which ignited the charges of all the barrels in rapid succession.* Weapons of this kind are to

* The trench warfare on the Western front has led to some rough-and-ready revivals of old methods. One of the illustrated papers recently published a photograph of what is practically an improvised

be found in many museums of arms, and we hear of their use in siege warfare in the fifteenth and sixteenth centuries.*

The contrivance was a clumsy one, and little advantage was to be gained from its use. A better device was a rude anticipation of the revolver principle. Armourers in the days of the Renaissance produced revolving firearms both for hand use and mounted on light carriages. The weak point of all such contrivances was that in the absence of machine tools it was difficult to make weapons in which the chambers of the revolver would all be truly bored, and the escape of gas between the revolving butt and the end of the barrel rapidly wore away the metal and made the weapon dangerous to use.

One may say that efficient machine guns could not be manufactured until machine tools had been invented, nor could they be perfected until the introduction of cartridges with a solid drawn copper or brass case made it possible to prevent dangerous escapes of explosive gases, and at the same time made it possible to clear the barrel for reloading by the complete ejection of the cartridge-case.

Nevertheless, many attempts were made to improve upon

---

revival of the organ gun. In a captured trench the French had taken a number of German Mausers and a quantity of ammunition. They fastened the rifles side by side in two rows, one above each other, on a wooden frame—twenty rifles in all—attached wires to the triggers, bringing all the wires to a central working point, where they could be drawn tight by pulling a wooden lever. Thus a volley of twenty shots could be fired by one pull of the lever to sweep the approach to the trench; and the Mauser being loaded with a clip of five cartridges, five successive volleys could thus be fired without recharging magazines, but each bolt must be opened and closed.

* In many European armies machine guns were first introduced as an auxiliary weapon for cavalry. It is therefore interesting to note that the first and only clearly recorded instance of the employment of the mediæval organ guns in the field was at the Battle of Piccardina, in 1467, when the Venetian General, Coleoni, brought them into action with the cavalry of his advanced guard. (Dittrich, "Machine Guns and their Employment," translated by Captain E. S. May, R.A., in the *Royal Artillery Journal,* 1889, p. 75.)

the old organ gun. The Artillery Museum at Woolwich has a weapon of this type dating from 1830; and an American form of the organ gun, known as the Requa battery, was mounted in one of the forts of Charleston during the American Civil War.

The first really practical machine gun was invented by Dr. Gatling, of Chicago, in 1862. The gun belonged to the non-automatic class. It may be described as a group of rifle barrels mounted, not side by side as in the old organ guns, but fixed at equal distances round a central axis, and mounted on a carriage like that of a field gun. Behind the six barrels the reloading mechanism was placed. This consisted of a cylinder containing the machinery which was worked by a crank-handle at the side, the cartridges being placed in a feeding apparatus on the top of the gun, and falling by their weight one by one into the gun as each shot was fired. On turning the handle, the loading apparatus and the whole system of barrels revolved round the central axis. A cartridge dropping into the gun was pushed forward into the barrel which at that moment was in the highest position with respect to the axis. As the barrel passed round to the lowest position, the cartridge was pushed home, and on reaching the lowest point it was fired. As the barrel passed up on the left side of the gun, the empty cartridge-case was extracted, and when the barrel again reached the upper position it was ready to receive another cartridge. Thus when the gun was in action one may say that at a given moment one of its barrels was firing, those on one side were being emptied ready for loading and the cartridges were being pushed home in those of the other three. The weight of the gun and its mounting prevented any recoil, and the water jacket surrounding the rifle-barrels over rather more than half their length prevented overheating.

A few of these guns were used during the Civil War. All kinds of experimental weapons were employed chiefly in the northern armies, but at the time very little attention was attracted in Europe by these incidents, and it was only long after that any records of the work done by the Gatling gun were published on this side of the Atlantic. The gun was never adopted during the war as an officially recognized weapon by the Washington War Department, and the curious fact is that when it was brought into action it was generally operated by an employee of the Gatling gun company, who took advantage of the fighting to demonstrate its powers.

The first machine gun that attracted general attention in Europe was the Montigny mitrailleuse, introduced into the French Army by the Emperor Napoleon III. on the eve of the war with Prussia in 1870. It was a very clumsy weapon, and the methods in which it was employed were generally so utterly devoid of any tactical knowledge that the result of the experiment was to bring machine guns in general into discredit for some years to come.

Mitraille is the French word for "grape-shot," and the new word *mitrailleuse* was adopted as the name of a weapon which would throw a shower of bullets producing the effect of a discharge of grape-shot at a longer range than that of this now obsolete kind of artillery projectile. One might perhaps translate the word *mitrailleuse* as "grape-shotter." In the French Army *mitrailleuse* is still used as the name for machine guns of all kinds, though it certainly is a very misleading description of their action. The Montigny mitrailleuse was secretly manufactured in the arsenal at Meudon in 1869, and the first months of 1870. Only the officers and men who were to work it were allowed to handle or even to see it. On the mobilization in July, 1870, the new guns were sent to the front carefully covered up with tarpaulins, and in the camps sentries were posted to prevent anyone from examining them. Meanwhile the newspapers published exaggerated descriptions of the terrible weapon that was to secure victory to France by mowing down whole battalions of Germans in a few minutes.

All this parade of secrecy was probably intended to impress public opinion in France with the idea that the Army had been provided with a terrible weapon that would prove a demoralizing surprise to the enemy. As a matter of fact, except as to minor details, the structure of the new gun was already known to many. Its general principle was not new. It appears to have been originally the invention of a Belgian officer, Captain Fafschamps, nearly twenty years before the Franco-German War. He made a rough model and elaborate drawings of his invention, and offered it to M. Montigny, a Belgian engineer and armourer, who had works at Fontaine l'Evêque, and a branch of his gun-making business at Brussels. This was in 1851, and a little later Montigny constructed some machine guns for the defence of the ditches of Belgian fortresses.* In the hands of the Montigny firm the gun was greatly improved, but it was not until 1869 that they succeeded in persuading the Emperor Napoleon III. to introduce it into the French Army.

In the summer of this same year, 1869, Major Fosbery contributed to the journal of the Royal United Institution a paper on "Mitrailleuses and their Place in the Wars of the Future." He stated that in the year before he had been employed by the Indian Government in reporting on the new gun, with a view to its adoption by the Indian Army, and that at the Belgian factory he had superintended the manufacture of a gun which was being tried at Woolwich. He mentioned that perhaps on this account his name had been associated with the invention, and he went on to say:

> I must, however, disclaim all share in the invention itself, which was entirely that of M. Montigny and Christophe, of Brussels. At the same time I consider that in urging on M. Montigny the adoption of Mr. Metford's rifling, and the hardened expanding projectile, instead of the ill-contrived, cannelured bullet formerly used with a

---

* Captain C. J. Tackels, "Armes de Guerre," Bruxelles, 1868, p. 180.

rifling, but ill-adapted to its object; in devising a system by which the cartridges are securely carried on service; and in changing the system of cartridge itself, I have been able to render the invention more suitable for the purposes for which it is intended.

Major Fosbery gives an illustration of the gun, and thus describes it:

The Montigny mitrailleur consists of an assemblage of barrels contained in a wrought-iron tube, mounted much on the same principle as an ordinary field gun, which, indeed, it somewhat resembles in form. To this a massive breech action is attached sliding between heavy iron plates. This is controlled by a jointed lever, and contains a simple contrivance for the separate and successive ignition of the cartridges. The cartridges are carried in steel plates perforated with holes, corresponding in number and position to the chambers of the barrel, of which, indeed, these holes form portions, being bored and finished with the same tools, and at the same time as the chambers themselves. These plates are about eleven millimetres in thickness, and when the cartridges (which are central fire) are dropped into them they stand out at right angles in proper position for introduction into the chambers of the gun. Grooves formed on the face of the breech-block receive the plate, which, being dropped into them, advances or retires with the breech-block itself.

A plate of cartridges being introduced into the gun, the gunner would depress the lever, which he holds in his left hand, the breech-block would advance, pushing each cartridge into its appropriate barrel, and finally becoming secured in its place beyond chance of accidental disturbance by recoil or otherwise. The contrivance which effects this is a very simple and perfect one. At the same time and by the same motion the whole weapon is set at full cock, ready for firing.

The gunner now quits the loading lever, and grasping the firing handle at the right side of the gun awaits the order to fire. One second of time is sufficient to give a complete revolution of the handle and discharge the whole of the thirty-seven barrels, of which the weapon consists, but each may be fired separately and at any intervals of time. When the last barrel has been discharged, he raises the loading lever, thereby opening the breech, and withdraws the empty cases by means of the plate, which now performs the office of an extractor, or rather of thirty-seven extractors in one. The plate is lifted from its groove, carrying with it the empty cases, and replaced by one filled with loaded cartridges, and the operation is repeated.

To open the gun, remove the empty plate, insert a full one, and reclose ready for firing takes somewhat less than four seconds. It has thus been found possible to fire this weapon twelve times per minute, throwing therefore 444 rifle shots in that time. This I believe to be the largest number ever yet thrown by a machine of like weight, though others pretend to the possession of greater powers in this particular.

The publication of such a detailed description in the summer of 1869 shows that there cannot have been much real secrecy about the all but identical French weapon when war broke out between France and Prussia in July, 1870. It appears also that officers of many armies had inspected the gun at Brussels, and one had been supplied to the Austrian Government for experimental trials. The Prussians knew all about it, and it came as no surprise to them.

The mitrailleuse was in all essentials the Belgian gun. Colonel de Reffye introduced some minor improvements in preparing the working drawings before the manufacture of the mitrailleuse was begun in the Meudon workshops. It weighed 1,800 kilogrammes, including the limber laden with 2,100 cartridges. A caisson carried 6,000 cartridges more. Six guns, six caissons, a travelling forge, and two baggage-waggons were given to each battery. Even at the time many officers of rank considered that whatever gain there might be in a surprise to the enemy—and we have seen how little hope of this there was—the advantage was outweighed by the serious drawback of giving a new weapon to the troops on the eve of the war without their having any opportunity for becoming familiar with it. Just in the same way on the eve of the war with Austria in 1859, the new rifled field guns were issued to the French artillery, with the result that during the course of the war the gunners had to find out the best way to handle them.

Whatever advantages it possessed were neutralized to a great extent by the mistaken use which the French Staff made of it. Perhaps because they looked so like field guns the new machine guns were placed in the hands of the artillery. Hitherto the French artillery had been organized for war in groups of three batteries, each of six guns. On the mobilization of 1870 a new organization was introduced. The artillery was still organized in groups of three batteries, but only two of these had field pieces, the third being armed with ten of the new mitrailleuses.

The events of the war proved that this was practically equivalent to reducing the fighting force of the French artillery by one-third, and giving them the poor compensation of adding to each group of batteries what was for all practical purposes a group of rifles. The extreme range of the mitrailleuse was that of the Chassepot rifle, with which the French infantry was armed—a little more than a thousand yards. The machine-gun batteries were therefore certain to be outranged by the enemy's Krupp guns, and they presented as good a target for shell fire as the French field gun, without having the power to reply effectively to this form of attack.

One may say, however, in mitigation of the blame for this fatal mistake that, strange as it may seem to us, many eminent soldiers of the time seem not to have grasped the elementary fact that artillery must do most of its work at ranges far beyond those of the rifle. The rifled gun was something new. It was only eleven years since it had been employed for the first time by Napoleon III. on the battlefields of Northern Italy. Smooth-bore guns were still to be found in many European armies, and batteries still came into action not infrequently side by side with the infantry. In his paper on machine guns, already referred to, we find Major Fosbery actually discussing the chances of a duel between a machine gun and a field gun, and suggesting that the result would be that if both were unloaded to begin with, the gun would never be loaded at all, and that "if both were loaded, and the first shot from the gun failed to smash the mitrailleur, the gun could not be loaded a second time; nor would a horse or a man belonging to it survive the first minute's practice from the weapon opposed to it." It is only fair to add that Major Fosbery granted that such a duel could only arise under exceptional circumstances, and that usually the field gun would come into action at a range beyond that of the machine gun.

The French, however, made the mistake of bringing their batteries of mitrailleuses into action side by side with the two batteries of field guns to which they were attached, with the result that they were again and again put out of action in a few minutes by the Prussian Krupp guns. Of course the new weapon was not entirely useless. If as a substitute for artillery it was doomed to failure, as a means of reinforcing the fire of infantry it was sometimes, though rarely, used with effect. It was first brought into action on August 2, 1870, in the small engagement which the French at the time dignified with the name of the battle of Saarbruck. Imaginative Parisian journalists described how the mitrailleuse mowed down the Prussians as a scythe levels the grass, or tore deep lanes in their advancing columns. The fight at Saarbruck was really a plucky rearguard action fought by a small Prussian detachment against the whole French division. There were only a few hundred infantry in action on the Prussian side, and they fought in wide extended skirmishing lines, offering the worst possible targets for machine-gun fire. The accounts of the terrible effects of the new weapon were newspaper fictions.

The first serious battle of the war showed the hopelessness of opposing machine guns in the open to artillery. At Wissemburg (August 4, 1870, about 10 a.m.) the artillery of the advanced guard of the German 11th corps opened fire on the French position at the château of Geisberg; a battery of French mitrailleuse galloped up to reply, and came into action on a knoll, marked by three poplars on the Geisberg height.

It was immediately marked down by the German artillery,

a Prussian shell blew up one of its ammunition waggons, mortally wounding General Douay, who commanded, and the battery was rapidly dismounted and had to be withdrawn from the fight.

At the battle of Spicheren (August 6, 1870), when General Valbrègue was forced to retire behind Styring, a battery of mitrailleuses of the Vergé division was placed near the road to close the gap left open at Styring in reply to the German guns. In a few minutes it had to withdraw.

But there were occasions when the mitrailleuse was used intelligently and with effect. One of these occasions was during the fighting on the French left and centre at the battle of Gravelotte (August 18, 1870). In the fighting on the left about Montigny-la-Grange, and in the centre at the Bois de la Cusse in front of Amanvillers, the French, instead of putting the mitrailleuses in line with their field guns and thus exposing them to long-range shell fire, brought them into action in small groups in their infantry firing line. The German official history of the war tells us what was the result. In the account of the fighting near Montigny-la-Grange we read that:

While the action was thus developing upon the principal line of battle of the 3rd division, three battalions of the advanced guard were having a partial engagement in the vicinity of Chautreune. Two companies of the 36th regiment had been led from Chautreune to the slope which rises from the east, but, as in the wood, they did not succeed in pushing further forward, for the open ground was entirely under the fire of the French infantry posted in the same wood and in the cluster of trees west of La Folie and Montigny-le-Grange. From this point notably, a battery of mitrailleuses swept the border northwest of the cluster of trees, and another battery from the south angle of this cluster held under its fire the clearing which separates it from the Bois des Genivaux. In a short time General von Blumenthal saw the impossibility of an attack upon La Folie. The position of the 36th regiment on the open ground afforded it very little shelter against the enemy's musketry and mitrailleuses, so its losses gradually reached a very high figure, while the French generally kept themselves defiladed, or outside the zone of action of the needle gun.

Of the fighting in the French centre we are told how the action of a battery of mitrailleuses led to the capture of the only guns lost by the Germans during the war. It will be seen that in this case the mitrailleuses were acting in combination with infantry at short range on the exposed flank of the German guns, which had been pushed forward into dangerous proximity to the French line under the false impression that this line did not extend to the north of Amanvillers:

> At this moment the artillery in position on the ridge to the south of the Bois de la Cusse was placed in an extremely critical situation. A battery of mitrailleuses had debouched in front of d'Amanvillers, and fired directly and with excellent range upon the extreme left of the line of Prussian artillery. This point was occupied by the fourth heavy battery already seriously injured by musketry. In a few minutes the fire of the mitrailleuses so decimated it that several officers, five chiefs of pieces, and forty men were disabled, and nearly all of the horses were killed or wounded. Such was the situation when suddenly large detachments of the enemy's infantry rose from the ravine in front of the ridge, and threw themselves with surprising swiftness upon the defenceless battery. Its chiefs, already wounded, succeeded with the few horses still untouched and after desperate efforts in getting two pieces back to the border of the wood, but the remainder of the pieces fell into the enemy's possession.

After Sedan not many machine-gun batteries were available for the French. Of those that they possessed about half were shut up in Metz with Bazaine's army, and most of what remained had been captured by the enemy at Sedan. Colonel de Reffye energetically pushed on the manufacture of new guns in workshops organized on the Loire as far down as Tarbes. Other guns were bought abroad. America sent guns made on the French system and some Gatlings; the trouble was that very few people knew how to handle the guns or had the remotest idea of their tactical use.

At the battle of Le Mans (January, 1871) Gatling guns were used successfully in the defence of the plateau of Anvours and the crossings of the River Huisne. The guns were used in a sensible way. They were kept out of sight of the enemy's artillery and used to check the German infantry by unexpected blasts of fire. Some machine guns were in trenches in the French centre.

Other instances might be quoted of the successful use of the mitrailleuse in battle. In all these cases it was employed to supplement or take the place of infantry fire. But at the time little was heard of these exceptional successes, a careful consideration of which would have pointed to the correct tactical principle of machine-gun work in war. What impressed the military opinion of Europe was the fact that in nine cases out of ten the gun had been a failure. But the criticism of the time was so *unintelligent* that a false deduction was made from this fact. All that it really proved was that it was a costly mistake to try to replace field guns with machine guns.

The only other machine gun used in the war was a weapon of the mitrailleuse type called "Feld," of which there were a few with the Bavarian army. A battery of four of these guns scored a success in the fighting at Coulommiers in the Loire campaign. It came into action with a French battery of artillery at short range (between 900 and 1,000 yards), and quickly silenced it, and then took part in the repulse of three French infantry attacks.*

The general result of the experience of 1870 was to discredit the machine gun as an arm for the battlefield. But at the same time it is generally recognized that its power of de-

* Report of the Swedish and Norwegian Commission on Machine Guns, 1872.—The weapon was the Feld machine gun, which had twenty-four barrels mounted in parallel rows. It was worked by a crank-handle firing about 300 shots per minute. The extreme range was 1,300-1,400 metres. It used the ammunition of the Bavarian infantry rifle. These cartridges were unsatisfactory for the purpose. The gun frequently jammed, the barrels also overheated easily, and then warped permanently in their frame, and had to be replaced. The failure of the Feld gun contributed to the dislike for machine guns generally which prevailed in the German Army for many years after the war of 1870.

livering a stream of bullets, or a series of volleys equal in effect to the fire of many rifles, made it a useful weapon for the defence of narrow passes and the flanking of fortifications. Naval men were also thinking of its advantages as a weapon for the fighting tops of warships, anti-torpedo boat armament, and for the equipment of boats and landing parties. Inventors were busy suggesting new types of gun, but for some time nothing was produced that could compete with the Gatling, seriously handicapped as this weapon was by the imperfect ammunition of the time.

In 1869 an English technical committee had considered the question of adopting some form of machine gun in the British service. It met again after the war and before the end of 1871 presented a report proposing that a certain number of Gatling guns should be purchased for trial in the Army and Navy. Two types of guns were adopted, a heavier gun using a ·65 cartridge for use in ships and coast defences, and a lighter gun with a ·45 cartridge for field service. The field service gun was intended chiefly for use in colonial war. The Gatling Gun Company sold a number of guns to Turkey, Egypt, Tunis, Morocco, China, and Japan. Most of these Governments were ready to do business with pushful agents, who had even less useful goods to dispose of than the Gatling gun. But the company's greatest success was with Russia. In 1871 General Gorloff was sent by the Czar's Government to the company's works at Hartford, with the mission of purchasing a number of guns constructed to use the Berdan cartridge of the Russian infantry rifle. Four hundred guns were delivered in a few months; most of these were distributed in the garrisons of European Russia, some of them being used for the auxiliary armament of fortresses, and others assigned for field service. One battery was provisionally attached to the cavalry for trial purposes. Forty-eight guns were sent to the Caucasus and twenty-four to Central Asia. The guns were all stamped with General Gorloff's name as that of the officer who had superintended their manufacture, with the result that for some years in the Russian Army machine guns were known as Gorloffs.

In the Central Asian campaigns the machine guns proved themselves highly effective weapons against the massed charges of the Turcoman cavalry. In the Khiva campaign on one occasion a section of two guns was the chief factor in the defeat of a dangerous attack made by an enormous force of the enemy in the grey of the morning. The advancing cavalry were barely visible, but the indistinct target was so large that it was impossible to miss it, and as the streams of bullets tore through the mass of horses and men, the Turcomans were seized with a panic and turned and fled, leaving heaps of their dead in front of the Russian line.

The new gun now manufactured in Russian arsenals, with some slight modifications of detail, was for some years a standard weapon of the Russian Army. It had been adopted some years before by the United States Navy, and a number of guns had also been attached to the troops serving in the Indian country and on the Mexican frontier. It had also been adopted in some of the cities of the United States by the police department for use in case of riot.

The Gatling gun, like all the earlier machine guns, suffered from the fact that at the time of its introduction into Europe the manufacturers of cartridges were not yet able to produce solid-drawn cases of absolutely uniform and reliable quality. The result was that with the earlier machine guns there were constant troubles from failure to extract the cartridge, and from jamming in the barrel or the mechanism. In the trials of the Gatlings at Vienna, though it was claimed that they were capable of a rate of fire of from 500-700 shots a minute, it was found that in actual practice the average was only 280 shots. "Almost every minute produced some interruption of the firing, owing to the sticking of the cartridge which failed to reach its proper barrel, or of a case which became jammed in the mechanism after leaving it." In our service the difficulty of making machine guns work effectively was increased by the fact that our cartridge was at the time one of the most defective in use in any army. It was the old boxer cartridge, which instead of the solid-drawn metal case had a case built up by rolling a thin brass plate round a mandril and attaching an iron base to it. Both with rifles and machine guns it was a by no means rare experience to find that the extractor tore off the iron base and left the chamber of the barrel blocked up with the brass part of the cartridge.

Nevertheless, with all these drawbacks, the advantages of a machine gun were so great that its adoption in some form was a necessity. In our service it was first adopted by the Navy, and Gatling guns were mounted in the tops of warships as a substitute for rifle fire, or on bulwark mountings to be used in defence against boat and torpedo attacks. But its first actual use in warfare was as an auxiliary weapon for naval landing parties.

Another machine gun, which in many ways resembled the Gatling, was the invention of Hotchkiss, an American engineer residing in France. The Hotchkiss machine gun, now in use in the French Army, is a weapon of a completely different type. The original Hotchkiss had, like the Gatling, a group of barrels mounted parallel to a central axis, round which these revolved during the process of loading and firing. But Hotchkiss made the barrels of his earlier guns of large calibre, so that they could throw small explosive shells instead of bullets, and the weapon was at first known as the revolver cannon. Of course the principle could also be applied to the firing of bullets, but it was the fact that it could throw shells which led to its early introduction into the French Navy as an armament for boats and pinnaces, and an auxiliary weapon for landing parties.

Before long two other types of rifle-calibre machine guns

began to come into use. One of these bore a resemblance to the mediæval organ guns. It was the invention of a Swedish engineer named Palmcrantz, and its official name was the Nordenfeldt-Palmcrantz gun, but it was generally known as the Nordenfeldt, taking its name from the Swedish banker who financed and managed the invention. The Nordenfeldt had a row of rifle barrels fixed side by side in a frame. At the back of the gun between the barrels and the breech action a smaller frame placed vertically carried the cartridges, which were placed above each other in five grooves, from which they slid down by their own weight into the loading apparatus. The gun was operated by a lever handle placed on the right. A gun of this type designed for naval use threw steel bullets of one-inch calibre, which would pierce the sides and boilers of a torpedo boat at moderate ranges. The other gun was the Gardner. In its original form it was like the Nordenfeldt, a gun of the organ type; but Gardner improved the reloading action, making it so rapid that he was able to diminish the number of barrels, and he thus produced first a two-barrel and then a single-barrel gun. There was a time when our Navy had in use a remarkable variety of machine guns, including Gatlings, Nordenfeldts, and Gardners of various calibres.

All the guns so far invented were operated by means of a lever or handle, which put and kept the loading, firing, ejecting, and reloading apparatus in action. But a new era of machine-gun construction began, when Mr. (now Sir) Hiram Maxim invented a gun which was really automatic—a gun which once started would go on reloading and firing itself so long as there were cartridges left in the feeding apparatus.

The Maxim gun is one of the most remarkable inventions ever made, if only because the first gun produced was practically perfect and did everything that its inventor claimed for it. There have been many modifications of the invention since it was first produced. Numbers of inventors have given their names to guns based on the Maxim principle.

But one may say that there has been no really essential improvement. The first gun ever made did as much as any of its successors have accomplished. So far the most rapid fire of machine guns had been at the rate of about 400 shots a minute. The first Maxim gun fired between 600 and 700 shots in the same time.

Sir Hiram Maxim has lately given a somewhat cynical account of the origin of its invention. He already had had a successful career as an inventor, and was in Europe as the agent of an American electrical company in the pioneer days of electric lighting. This is his account of how he first came to think practically of work in the new department of machine-gun making:

> In 1881 I visited the Electrical Exhibition in Paris, and was made a Chevalier of the Legion of Honour on account of some electrical and chemical work that I had done; and about a year later I was in Vienna, where I met an American Jew whom I had known in the States. He said: "Hang your chemistry and electricity! If you wish to make a pile of money, invent something that will enable these Europeans to cut each other's throats with greater facility."
>
> This made me think of the time when I was only about fourteen years of age, and was making drawings for my father of a hand-worked machine gun. I also thought of the powerful kick I got the first time I fired a U.S. military rifle. On my return to Paris I made a very highly finished drawing of an automatic rifle. Happening to meet a Scotchman in Paris whom I had known in the States, I showed him my drawings. He invited me to come to London. I did so, and shortly after I started an experimental shop at 57d, Hatton Garden.*

In the more detailed account of the process of his invention, which Sir Hiram Maxim gives us in his autobiography,† he makes it clear that after the first suggestion had been made to him in Paris there was a gradual change of plans. The first sketch of the gun reproduced at p. 158 of the Autobiography is a drawing of a repeating rifle to be fired from the shoulder. When he returned to London and started his

* Letter to the Editor of the London *Star*, July 23, 1915.

† "My Life," by Sir Hiram S. Maxim. London: Methuen, 1915.

drawing-office and workshop at Hatton Garden, he designed not a repeating rifle but a machine gun to be fired from a tripod.

Patents were taken out not only for the device which was finally adopted, but for several possible variations of it.* Gunmaking was new work for Sir Hiram Maxim, but he had the great advantage that he was not only a first-class mechanic but had also plenty of money behind him, and could afford to spend it freely in carrying out his ideas. The new gun was not shown to anyone outside his immediate group of helpers until it was practically perfect.

The principle adopted was to use a single gun barrel and to mount it so that after firing each shot it recoiled in its mounting through a short distance, elongating a spring which brought it back to firing position at the end of the recoil, the double movement backwards and forwards at the same time working the mechanism which ejected the fired cartridge and loaded up another in its place. Hitherto all machine guns had been loaded either by a mechanical feed worked by the firer, or by the weight of the cartridges dropping one by one into position from some kind of hopper or magazine above the gun. Maxim devised a new kind of feed, the cartridges being held in clips in a belt which the action of the

* The first of the Maxim patents is dated July 16, 1883 (No. 3493). It is for "an invention of improvements in machine or battery guns, and in cartridges for the same and other firearms." In the No. 3493, the feed-block was in the bottom of body. There are further patents, dated 1884, January 3 (No. 606) and February 23 (No. 3844), for machine guns. There is a patent, May 23 of the same year (No. 9407), for a cartridge feed, and, October 2 (No. 13113), for the lubrication of machine guns. There are further patents, dated January 29, 1885 (No. 1307), and July 8, 1885 (No. 8281), this last being the first patent in which the machine gun assumed much the present Maxim form. There are further patents in 1890, 1892, 1894, and 1895 in the names of Sir H. S. Maxim or jointly with his assistant, L. Silverman. The 1895 patent (No. 5864), dated March 20, describes the form of lock now in use.

gun worked forward from right to left, so as to bring a cartridge up to the loading position after each shot was fired.

One of the critical questions to be settled was the distance through which the barrel was to be allowed to recoil to give the best result. This was determined by constructing an apparatus in which the parts could be adjusted to allow of various lengths of recoil. The very first trial gave a satisfactory result, six cartridges being fired off in about half a second. "I was delighted," writes Sir Hiram Maxim. "I saw certain success ahead, so I worked day and night on my drawings until they were finished, and went into the shop and worked myself until I had made the gun. It was finished in due time, and on trying it with a belt of cartridges I found that it fired rather more than ten a second."*

Several guns were made. So far the work had been carried on privately, but now the newspapers began to talk of the automatic machine gun which would load and fire itself by energy derived from the recoil and send out a stream of more than 600 bullets in a minute from a single barrel. There were almost daily demonstrations before distinguished visitors in the Hatton Garden workshop. The first to come was Sir Donald Currie, and a few days later the Duke of Cambridge and the Prince of Wales, afterwards King Edward VII., came to see the gun firing. Next came Lord Wolseley and a group of War Office officials. The gun was shown at the Inventions Exhibition at South Kensington in the summer of 1885. Government trials were arranged, and contracts were made for the manufacture of the gun at works belonging to Messrs. Vickers at Crayford, in Kent.

Among those who took the keenest interest in the new gun was Lord Wolseley. Discussing the design with the inventor, he suggested that a larger cartridge and barrel should be used, giving a longer range than that of the infantry rifle, and he asked Mr. Maxim if he could not make a long-range gun with

* "My Life," p. 170.

a projectile that would penetrate barricades and other obstructions. Sir Hiram told him that such a gun would not be so effective as a smaller gun in stopping a mad rush of savages, because it would not fire so rapidly, and, after all, nothing bigger than the service bullet was needed to kill a man. Lord Wolseley then suggested that a gun might be devised which would throw a heavy long-range projectile and yet be able to stop a rush as effectively as the small gun, and this led to a new invention, which Sir Hiram thus describes:

> After thinking the matter over for some time, I designed a gun, made, and patented it. It had a bore of three-quarters of an inch, a large powder charge, and a very peculiar kind of projectile, with a hardened steel core. The projectile was made up of several lead segments arranged around the central steel core, and the whole held together by rings of lead hardened with tin. In firing at long range the projectile acted the same as any other projectile. On striking the steel plate the hardened steel core would easily pass through the sides of the caissons, and in case of a mad rush at very short range it was only necessary to move a handle to bring four cutters into position in front of the muzzle that would cut and weaken the lead rings sufficiently to allow the projectiles to fly apart, and to act very like buck shot out of a large shot-gun.*

We have no record of any use of this gun on active service, but the designing of a long-range gun led to a new development. In order to enable the gunner to see where his shots were going and correct the range, a small shell would be more useful than a big bullet. This led to Sir Hiram Maxim designing a gun of the type afterwards popularly known as the "pom-pom," a large calibre Maxim gun, throwing a shell weighing a little more than a pound—half a kilogramme—the lightest weight allowed for explosive projectiles under the terms of the Geneva Convention.

But these subsidiary inventions were of far less importance than that of the original automatic gun sending out a stream of bullets at the rate of ten a second. It was one of the few

* "My Life," p. 183. This invention was covered by a patent for "muzzle cutters," November 17, 1885 (No. 14047).

inventions that had no wearisome time of waiting between the stage of experimental trials and business results. Sir Hiram Maxim travelled all over Europe showing it to the officials of various Governments. At Berlin the Kaiser saw the gun fired and gave his opinion of it in the emphatic words: "This is the only machine gun." But even the Kaiser was not always able to impose his opinions upon the chiefs of the German Army, and for some years the General Staff at Berlin hesitated to adopt machine guns of any kind. It is a curious instance of the effect produced on the minds of practical men by a strong impression really based on insufficient data. For years, when there was a question of machine guns, the German Generals would dismiss the matter with some such remark as: "We know all about machine guns. We saw in the war of 1870 what a mess the French made of them, and we would rather have the same number of field guns instead of them." The experiences of 1870 seem in other cases also to have led to the false line of reasoning, that the machine gun must be judged by comparison with the field gun.

With other Governments Sir Hiram Maxim was more successful. It is interesting to note that the Chinese envoy, the celebrated Li Hung Chang, when he saw the gun fired at Crayford, was greatly impressed, but raised an unexpected objection to its adoption by his own country. Several belts of cartridges were fired off, and, amongst other experiments, a tree some hundreds of yards away was cut down by the sweeping fire of the gun. Li Hung Chang asked what was the cost of the cartridges fired off in a minute, and was told it was nearly £30. "That gun won't do for us," he said; "it is much too expensive for China."

The original Maxim patents have long since expired, and the guns now manufactured by the Vickers Company are covered by patents on various improvements of detail. As already noted, the principle of the gun has been made the basis of other inventions, and it may be said that the greater number of types of machine guns are practically variations of the original Maxim design.

But there is another class of machine guns, in which the motive power is derived, not from the recoil, but from a portion of the gas generated by the explosion of the cartridge being allowed to escape through a valve in the barrel before the bullet has left the muzzle, the gas thus escaping being made to work a piston or similar device, which actuates the extracting and reloading and firing mechanism. Some of these forms of machine guns will be noted later in describing the armament of the European armies of to-day. Here it may be said that, whatever their merits may be, few of them have given results equal to the guns built on the recoil principle, and none of them have given better results. The Maxim in its various forms is still the standard machine gun.

In tracing the evolution of the machine gun, we have mentioned the first form of the Hotchkiss, the revolver cannon—a gun with a group of barrels somewhat resembling the Gatling type. The machine gun, now known as the Hotchkiss, which is the weapon of the French Army, is a very different kind of gun. It is automatic, but belongs to the second or gas-engine type of guns. The gun is worked by a portion of the gas escaping through an automatic valve at about one-third of the length of the barrel from the muzzle, and working a piston in a tube fixed underneath the barrel. There is no water jacket. The gun is air-cooled, with the help of a number of projecting flanges on the barrel in the region of the powder chamber. It is made in two forms—a heavier form carried on a carriage or pack saddle and fired from a tripod, and a lighter form usually carried on a pack saddle, but handy enough to be carried by a man if necessary, and almost light enough to be classed with repeating rifles.

The German Army uses the Maxim gun. It is manufactured for the Army by the Deutscher Waffen und Munitions Company, near Berlin. It had already been introduced into the Navy, when in 1899 the first Maxims were issued to the Army. They were in the first instance given to the Jaeger battalion attached to the cavalry corps, and were first brought into the field at the manœuvres in that year in South Germany. The guns were organized in groups of four, a group being attached to each battalion, and all four guns being carried in a waggon drawn by two horses. At the same time, at his own expense, the Kaiser gave a Maxim gun to each of the Dragoon regiments of the Guards. These first trials were largely experimental, with a view to working out an appropriate organization for the new weapon and studying its tactical use. In 1900 batteries of Maxims were attached to the cavalry divisions, and at the manœuvres in the following year there was a cyclist company with a Maxim battery attached to one of the cavalry divisions. From this date forward there was a steady increase in the number of Maxims and Maxim detachments. The guns were attached to both cavalry and infantry brigades, and at the beginning of the present war the Germans took the field with tens of thousands of Maxims.

These facts are noted here, but a fuller account of the use of the Maxim gun in the German Army will be found in a later chapter. Before describing in greater detail the machine-gun armament of the various armies, it is interesting to note that, like our own, and the German Army's, those of the greater majority of the Powers have adopted a gun of the Maxim type. In 1905 Captain von Braun, a well-known German student of this subject, in his book "Das Maxim Maschinengewehr und seine Verwendung," gave the following statistics of the armies and fleets using the various machine-gun systems:

System.	Armies.	Fleets.
Maxim	19	21
Hotchkiss	4	2
Other systems	5	1

The Maxim system was introduced into our own Army in 1891, and rapidly displaced the various types of machine guns previously in use—Gatlings, Nordenfeldts, and Gardners. Machine guns, as already mentioned, had been adopted for the Navy before they were introduced into the Army. Some of the first cases of their employment on active service were with naval brigades landed to co-operate with our troops in minor wars in various parts of the world. The machine gun's power of stopping a wild rush of savage warriors led to its early employment in Indian and colonial wars, but for a long time there was outside a small group of enthusiasts a strong prejudice in Army circles against the idea of its being used in regular warfare. Possibly, as in Germany, the false conclusions drawn from the events of the Franco-Prussian War had much to do with this. Here, as in so many other cases, the pioneer work that eventually secured the adoption of a new policy was done by the old Volunteer Army, now the Territorials. While the Regulars had to await the decision of the War Office on any new departure, the Volunteers were freer to adopt new methods at their own expense. In the face of much adverse criticism, Colonel Alt, of the Central London Rangers (then the 22nd Middlesex), brought into the field at manœuvres a couple of Nordenfeldt guns. A little later the 4th Volunteer battalion of the Royal Fusiliers had a machine-gun section, and the cyclist battalion, the old 26th Middlesex, invented a light carriage for the Maxim, which was drawn as a trailer by a couple of cyclists, kept up with the battalion over all kinds of roads, and could be dragged by hand wherever a firing line could go. Machine guns were also adopted by other Volunteer battalions in the provinces. These experiments were of the utmost use when the question arose of attaching machine guns to the infantry battalions and cavalry regiments of the Army.

Such is in brief outline the story of the evolution of the machine gun, and its entrance into the armies of Europe as a recognized weapon. It will now be of interest to say something of experiences on active service with the machine gun in those early years.

CHAPTER III

MACHINE GUNS IN BATTLE

AS we have seen, the first European army which added machine guns to its armament, after 1870, was that of Russia. The guns had done good service in expeditions against the Turcomans in Central Asia, and when the war with Turkey began in 1877 a number of batteries of improved Gatling guns were sent to the front. At the siege of Plevna they were used for defending the Russian entrenchments, and on several occasions they proved effective in stopping sorties of the Turkish garrison. For this purpose the gun was placed in position at twilight, and marks were set up to show in what direction it should be trained, in order to sweep with its fire bridges or other narrow lines of approach.

Gatling guns were also used by the Peruvians in the war with Chili in 1879 for the defence of entrenchments. In England machine guns were first adopted for naval purposes. Messrs. Thorneycroft, whose works were then at Chiswick, had specialized as builders of powerfully engined and rapid steam launches. Out of the river launch originally built for pleasure purposes, they evolved first small torpedo boats, carrying a spar torpedo, and then a larger type of boat, carrying a torpedo-tube on deck. The first torpedo boat for the British Navy was built at Chiswick in 1876.

With the general adoption of this type of boat by the European navies, the problem arose of how best to meet this new form of attack. The torpedo boat was built of very thin steel plating in order to economize as much weight as possible, and be able to put it into the engines. It was suggested that the best weapon against such boats would be a machine gun firing either ordinary bullets or hard steel projectiles or small shells. Quick-firing artillery was not yet in existence. The period between the time when a torpedo boat approaching under cover of darkness would be discovered and the moment at which she would be near enough to fire her torpedo would be very brief, and although a shell from one of the lighter guns on a battleship or cruiser would break the torpedo boat if it made a fair hit, the chances of such a hit on the small rapidly moving target presented by the enemy would be very slight. The machine gun, with its power of sending out a continuous stream of small projectiles, was obviously a much more reliable defence than the ordinary armament of the warship.

In May, 1880, experiments were made at Portsmouth to determine the stopping power of the machine guns then available against a torpedo boat. A gunboat, the *Medway*, had two machine guns—a Nordenfeldt and a Hotchkiss revolving gun—mounted on her forecastle. While firing, the gunboat was kept under steam at full speed, starting from a range of 1,500 yards, and running up to within 600 yards of the targets. The targets were full-size models of torpedo boats, with dummy machinery inside of them. They were placed sometimes broadside, at others bows on to the gunboat. The machine guns' fire did such damage that it was considered that in actual warfare the torpedo boats would have been completely disabled. The Nordenfeldt riddled the target; five of its steel bullets hit the piston rods of the boat's dummy engine, and it was believed that any one of these shots with the engine running at high speed would have thrown the machinery out of gear. The Hotchkiss gun also made good practice. Its shells, weighing about a pound each, had a bursting charge of rather less than an ounce of gunpowder. Numbers of them burst inside the target boat, the fragments made hits all over the structure, and in several cases fragments of the shell passed out through the steel plating on the opposite side.

These experiments decided the question of the adoption of the machine gun as an auxiliary armament for warships.

A variety of guns were purchased, including ten-barrelled Gatlings of ·45-inch and ·65-inch calibre; Gardner guns, mostly with five or two barrels, but a few of them with a single barrel, all of the ·45-inch calibre; and Nordenfeldt guns of two classes—a five-barrelled gun, throwing a ·45-inch bullet, and a gun by the same maker which was known as the Nordenfeldt anti-torpedo boat gun. It had four or two barrels, of one-inch calibre, and fired hard steel bullets. (*Naval Annual*, 1886.)*

At the bombardment of Alexandria in the summer of 1882 machine guns were mounted in the tops of the ships, in order to bring a plunging fire upon the Egyptian batteries, but the results were disappointing, probably on account of smoke obscuring the view; the ranging was bad, and an examination of the batteries after the bombardment showed very few marks of machine-gun bullets. During the attack the gunners in the tops of the ships were wrapped in a fog of powder-smoke from the big guns lower down, and the machine guns themselves produced another dense cloud round the tops. One may say that smokeless powder is almost a necessity for the thoroughly efficient working of any kind of quick-firing gun, whether cannon or machine gun.

The first important land operations in which machine guns were used with a British force were those of the Zulu War of 1879. The guns were Gatlings, which were already in use in the Navy, but had not yet been recognized as a Regular Army

* In a lecture on machine guns delivered at the Royal United Service Institution in June, 1883, Lord Charles Beresford said that by the end of the financial year, March 31, 1884, the Navy would possess 565 Nordenfeldt guns, throwing a solid steel inch bullet, chiefly for repelling torpedo attacks. There would also be 350 Gardner guns, and 142 Gatlings.

weapon. Of the use of the machine guns in this campaign, Lord Chelmsford, who was in command, wrote as follows:

On the advance to the relief of Ekowe, two Gatling guns accompanied the column, and at the battle of Ginginhlovo did considerable execution amongst the Zulus *at the opening of their attack,* which commenced on the north side of our position. The Zulus very soon, however, worked round to the west and south side of our laager, and the Gatlings were not in action therefore for any length of time.

At Ulundi we also had two Gatlings in the centre of the front face of our square. They jammed several times in the action, but when in work proved a very valuable addition to the strength of our defence on that flank. Machine guns are, I consider, most valuable weapons for expeditions such as that which we had to undertake in Zululand, where the odds against us must necessarily be great, and where it is necessary to leave small detachments in charge of posts along the line of communications. The Gatlings, however, required too much care in firing, and could not be entrusted to any but skilled manipulators. If a machine gun can be invented that may safely be entrusted to infantry soldiers to work, and could be fired very much as one grinds an organ, I am satisfied of its great value. They should, however, in my opinion, not be attached to artillery, but should be considered as essentially an infantry weapon, and should be worked by infantry soldiers. So utilized, they might, I feel sure, be used most effectively not only in defence, but *in covering* the last stage of an infantry attack upon a position, where the troops have at last to cease firing and endeavour to get home with the bayonet.*

In the fighting near Suakin with the Dervishes, and on the Upper Nile, 1884-5, machine guns—Gardners—were employed with the naval brigades landed to co-operate with the troops. Lord Charles Beresford, who commanded the naval brigade in the Nile Expedition, summed up the experiences with machine guns in these two campaigns in an interesting lecture delivered at the Royal United Service Institution. He pointed out from his own experiences, and those of other naval officers, that the machine guns would have been more efficient if they had been better mounted. The progress of the new weapon was still being hampered by the traditional idea that it was a new kind

* *R.U.S.I. Journal,* 1885, p. 947.

of field gun. It was mounted on an artillery carriage, and hooked on to a limber which carried the ammunition. Most of the guns were mounted exactly as if they had been field guns, being fixed by means of trunnions between the wheels and a very little above the level of the axle. The result was that in order to train the gun to right or left the trail had to be moved round, and this made any fine adjustment of the aim a difficult operation. As Commander Rolfe, who served with the naval brigade in the Suakin operations, pointed out, it was really a question of delicately adjusted rifle fire, and what was wanted was a screw for training the gun to right or left, instead of having to make a rough adjustment by pulling the trail to one side or the other.

The sailor is proverbially a handy man, and the chief engineer of one of the ships improvised a mounting for two of the Gardners, which remedied this defect. The guns were mounted on seven-pounder field carriages, but were carried up above the level of the wheels on a kind of pivot arrangement, by which they could be swung round through more than half a circle without moving the gun-carriage. This was found to be a great gain, and it had the further advantage of bringing the gun up level with the shoulder of the man who was firing it. But, as Lord Charles Beresford pointed out, the whole arrangement of mounting the gun as if it were a cannon was a bad one. Limbering and unlimbering, he said, "does away with much of its utility as a machine gun. The limber adds to its weight, and it is not ready for instant action." He read a letter from Captain Wilson, V.C. (now Admiral Sir Arthur Wilson), in which further disadvantages of the gun-carriage arrangement were pointed out. He remarked that the operation of coming into action by swinging round the gun and unlimbering required a certain amount of space and caused some inevitable confusion at a moment when the men were forming square to resist an enemy's rush, and there was even for a few moments a dangerous gap, which had to be adjusted when the gun was in position. But once the guns were in action there was an obvious advantage in the easy control of the concentrated rifle fire which they supplied. Thus, describing one of the Soudan battles, Lord Charles Beresford says:

> I was much impressed with the ease with which the fire of these guns could be directed or controlled by the officer in command, in comparison with the rifle fire of both soldiers and bluejackets. The men were at the time very excited, the noise and general confusion preventing orders from being heard. Mounted officers rode furiously up and down the line with little effect, while the bugles almost continually sounding "cease firing" seemed only to add to the noise. But the machine guns were under perfect control, orders quietly given to "Search out that clump of bush!" "Keep your gun bearing on that corner of the wall!" or "Cease firing till they show again!" were carried out with the greatest regularity.

With both Gatlings and Gardners the defective cartridges of the time—the old Boxer pattern—caused some anxious moments when firing had to be stopped and a broken or jammed cartridge-case cleared away. But even the best of machine guns can be hampered by defective ammunition. Perfection in the cartridge is as important as perfection in the gun.

During the campaign against the Dacoit bands in Burmah, after the occupation of Mandalay, a machine-gun battery was organized and commanded by Captain W. N. Lloyd, R.A. The guns available were four two-barrel Gardners, handed over by the Navy, and originally intended for use by the naval brigade. In the jungle fighting wheel transport was out of the question. Captain Lloyd arranged for pack transport for guns and ammunition. About 16,000 rounds were carried with the battery. This was a sufficient supply, as the skirmishes with the Dacoits were generally of brief duration. It was found that the guns could be taken from the pack saddles, mounted on the tripods, and brought into action in about half a minute. The enemy seldom made a serious stand, but, where they did so, the four Gardner guns proved very useful. In the accounts of the work of his battery, contributed to the *Royal Artillery Journal* (1886), Captain Lloyd says:

When the Dacoits opposed our advance by clinging to the jungle in our front, their position, never extensive, would be quickly searched out by the field guns. Again, their value would be appreciated in storming stockades, some of which were bullet-proof, others not. In the latter case the guns, having a range of 2,000 yards, could keep up a stream of bullets (provided the country admitted of it) out of the enemy's fire, while in the former case, the stockade, having been sufficiently pounded by artillery fire to enable our men to rush it, machine guns would take up a position bearing on the line of the enemy's retreat. In like manner they would be utilized in the attack on Dacoit villages. For passive defence the Gardner guns might be employed in our military posts, which have in several instances been hard pressed by Dacoits. Moreover, the power of these guns for counter-attack, as well as for passive defence, cannot fail to be recognized.

The Maxim gun was first used by British troops in campaigns on the North-West Frontier. Before this it had been employed with remarkable effect by colonial forces in the Matabele War of 1893. The troops engaged against the Matabele were the armed police of the Chartered Company and volunteers enlisted for the campaign. They were heavily outnumbered by the enemy, who invariably attacked with all the reckless courage that the Zulu tribes have so often displayed in war. The Rhodesians usually adopted the tactics of acting at the outset on the defensive and provoking the Matabele to charge. The handful of white men temporarily fortified themselves by forming a waggon laager with the Maxim guns at the angles. The reports of the time tell again and again of the failure of the Matabele charge in the face of the fire of the machine guns. Thus, in an account of the battle near Fort Victoria at the end of October, 1893, between Major Forbes' column and a large Matabele army commanded by Lobengula in person, we are told that:

Most of the Matabele had probably never seen a machine gun in their lives, and had but a dim idea of the effects of concentrated rifle fire. Their trust was in their spears, for in all their rude experience of warfare they had never known an enemy able to withstand them. Even when they found their mistake, they had the heroism to regard it as only a momentary error in their calculations. They retired in perfect order, and re-formed for a second rush. In how many European armies could the men who had survived one shower from modern artillery come forward to try their luck again? These savages were equal to the attempt, and equal, too, to the deliberate design of bettering their luck by looking for a weak place in the laager. Once more the Maxims swept them down in the dense masses of their concentration, and once more they retired. It seems incredible that they should have mustered for still another attack, yet this actually happened. But by this time they had reached the limits of human endurance. They came as men foredoomed to failure, and those who were left of them went back a mere rabble rout.*

From another account it appears that in the first and second charge some of the Matabele got so close up to the laager that they were able to throw their spears into it. There can be little doubt that it was the Maxim gun that saved the small white force from destruction against such determined rushes of thousands of fierce fighting men, the number of rifles engaged was not sufficient to break up the charge. It was the stream of bullets from the machine gun tearing lanes of dead through the enemy's masses that broke the rush. Of course it must be remembered that the primitive tactics of the Zulus made them an exceptionally vulnerable target for machine-gun fire. The guns in action were four Maxims, a Gardner, and a Nordenfeldt. But the fire of the Maxims was the heaviest and the most rapid, and did most of the work.

It would be easy to multiply accounts of actions in the Matabele War in which the machine gun played a decisive part.

In the spring of 1895, in the march to the relief of Chitral, Maxim guns were in action with remarkable effect against the

* London *Daily News*, November 3, 1893.

rushes of the Ghazis at the Malakand Pass, and were also found extremely useful in the attack in driving the enemy from their sangars, or stone breastworks on the hillside. In the fight near Gumbat in the middle of April, a rush of the hill men was stopped by a single Maxim being brought into action against their flank as they advanced.

Summing up the experiences of the expedition, a military correspondent of the London *Globe* wrote on May 15, 1895:

> I am glad to find that the Maxim guns have done so well in the Chitral Expedition. By a stroke of good luck six ·303 Maxims reached India a few weeks before the operations began. The military authorities were thus enabled to issue three to the troops, and to send up an unlimited supply of cordite ammunition of the same kind as that used in the Lee-Metford rifle. The guns were sent out from England on wheeled carriages, but each regiment which received one fitted it for mule transport. The guns will probably be used in all future expeditions, and will, I trust, be taken up with increased interest at home. Talking of machine guns reminds me that some officers still advocate the Gardner for use with cavalry. Yet the Maxim with tripod does not weigh more than 40 lbs., while the Gardner weighs 100 lbs.; and in regard to rapidity of fire and accuracy the Maxim has much the best of it, and is in all respects a far superior weapon.

In the official despatch describing the operations, published in the *London Gazette* of June 14, 1895, the following reference was made to the services rendered by the Maxim guns in the fight for the Malakand Pass:

> Sangar after sangar was obstinately held. Each sangar as it was rushed coming at once under the fire of the one above it, and here I may note the admirable service done by the artillery and Maxim guns. Several attempts were made by the enemy to concentrate from above and hold lower sangars and positions, but all such attempts were frustrated by the admirable practice of the mountain batteries and Maxim guns over the heads of our advancing infantry.

It may be said that these experiences of the value of even a few machine guns in mountain warfare settled the question of the general adoption of the Maxim gun in the British Army. In the following year the campaigns for the reconquest of the Soudan began, when in 1896 Major-General, afterwards Lord Kitchener, the Sirdar of the Egyptian Army, organized the small force which advanced into the Dongola province. He was anxious to carry the operation through, at least in its initial stages, with Egyptian troops only. But he sent to the front the machine-gun sections of two of the British battalions, then in Lower Egypt—the Staffordshires and the Connaught Rangers. Each battalion had two guns in its machine-gun section, and these were organized as a battery of four guns, the guns being mounted on wheeled carriages, but having also mule pack-transport. A special precaution had to be taken to prevent the mechanism of the gun being jammed by the desert sand, which forms almost a fog on windy days in the Soudan. Each gun had a silk cover in which it was kept wrapped up until it was to be brought into action.

In the first fight at Ferkeh (June 7, 1896), the machine-gun battery, working in sections of two guns each, was only in action for a few minutes, but in this time it broke up the only attempts to charge made by the Dervishes.

In the following year the Maxim battery received two additional guns: 5,000 rounds were carried with each gun, and a reserve of 20,000 cartridges followed, carried on mules or camels. In the battle of the Atbara the machine guns were boldly pushed out to bring a flank fire to bear upon the enemy. In the final battle at Omdurman the Maxim battery was on the right front of the Zereba between two of the Egyptian brigades. Other machine guns were in action with the British division on the left of the line. Major von Tiedemann, the German military attaché, tells how, when the great Dervish attack began, he rode away from the Staff in order to watch the effect of the fire of the Maxim battery on the right. He says that the gunners did not get the range at once, but as soon as they found it the enemy went down in heaps, and it was evident that the six Maxims were doing a large share of the work of repelling the Dervish rush. They were firing ·303 cartridges, with smokeless powder, and besides the rapidity of their fire, they had the advantage that their weapon had a longer range and a flatter trajectory than that of the Martini rifles, with which the Egyptian troops to right and left of them were armed.

In the first year of the Soudan campaign, while the Sirdar was advancing into the Dongola province, there was a widespread native rising in Rhodesia. Here again the Maxim guns in the hands of Colonial troops and British regulars proved formidable weapons in fights against superior numbers of the enemy. But so far the Maxim gun had been used almost entirely against the crude tactics of savage or half-civilized opponents. It was still a question what it would do in warfare between regular armies. Attempts to solve the problem had been made at peace manœuvres, but manœuvre results are largely theoretical. There are no bullets in the guns, and imagination has to play a considerable part in estimating results.

There was a battery of four Gatling guns with General Shafter's army in the Cuban campaign of 1898. It was improvised after the landing, and the gunners had only a few days' training. Nevertheless, the guns did some effective work. At a critical moment of the battle before Santiago, the guns were pushed into the firing-line and opened fire upon the Spaniards on San Juan Hill. Three guns fired 6,000 rounds each, and Lieutenant Parker, who commanded the battery, claims that it was this heavy machine-gun fire which made it possible for the assault on the hill to be carried through successfully in a few minutes. As soon as the hill was taken, the guns were rushed up to it, and materially contributed to its defence against the Spanish counter-attacks.*

In the South African War machine guns were used on both sides. The guns were usually Maxims of two kinds—the

* "Tactical Organization and Uses of Machine Guns in the Field," by John H. Parker, Lieutenant 13th U.S. Infantry, p. 41.

rifle-calibre gun, used by both British and Boers, and a gun of larger calibre, which was used only by the Boers in the first stage of the war, but a few of which were later on used on the British side also. This larger calibre machine gun was popularly known as the "pom-pom." It was a Maxim, with a heavy barrel of 1·457-inch calibre, designed to throw small steel shells, weighing one pound, the smallest weight of an explosive projectile permitted by the Geneva Convention. The small shells were fired from belts, each of which held twenty-five projectiles. The gun had been offered to the British Government before the war, but had been refused. The Transvaal Government, however, purchased a considerable number of these big Maxims, regarding them as a cheap and handy form of light artillery.

The experience of our own men under pom-pom fire was that the little shells seldom did any serious damage, but it was said that the succession of sharp explosions of ten shells bursting one after another on nearly the same spot got on men's nerves even more seriously than rifle or artillery fire. Towards the end of the war pom-poms captured from the enemy, and a few bought in England, were employed with our flying columns. For awhile there was a talk of batteries of this weapon being permanently attached to our cavalry brigades, but the idea was subsequently rejected, and the decision was sound. The pom-pom had the serious drawback that it could not deliver the rapid sustained fire of the smaller calibre machine gun. Its shells were, comparatively speaking, ineffective, and yet it required a carriage nearly as heavy and as prominent a target as that of the field-piece. Quick-firing artillery was being adopted by all the armies of Europe, and the pom-pom was really a rather feeble kind of quick-firing gun. It was obvious that it would be better to have the more powerful weapon instead of a device which was a kind of compromise between the rifle and the cannon.

The records of the South African War show that machine

guns were used in comparatively small numbers, usually as supplementary weapons with an infantry or cavalry unit. We find no instance of their being grouped in batteries or handled with much tactical initiative. On the Boer side we hear of a single Maxim sometimes being brought into action with the field guns.* The Boer artillery was very weak in numbers, though the guns were mostly of a better type than our own, and they were handled with remarkable intelligence. It is possible that the association of the machine gun with the artillery was based on the sound tactical idea of the machine-gun section serving as an escort to the guns.

In the defence of entrenched positions machine guns proved very useful. In the narrative of the defence of Mafeking in the first sortie against the Boers on October 14, 1899, the armoured train, which took part in the operations, was armed with two machine guns, a Maxim and a Hotchkiss. In the advance on Pretoria there was a battery of pom-poms with the artillery and about fifty machine guns, distributed singly among the various regiments engaged. In the later stage of the war, when flying columns were used to hunt down the Boer guerillas, there were generally three or four Maxims with each detachment; but very rarely in the history of the Boer War do we find any clear instances of machine guns being used with marked or decisive effect.

The first war between regular armies in which machine guns were used in large numbers on both sides was the war between Russia and Japan of 1904-5. The Russians had the Maxim gun, the Japanese the Hotchkiss gun, of much the same pattern as that now used by the French Army.

On both sides the guns were grouped in batteries of six or eight pieces. The machine gun was thus used as a special arm. With the Russians, the battery was made up of eight

* In the despatches and the British official "History of the War," when a "Vickers-Maxim" gun is mentioned as in action on the Boer side the large calibre Maxim (or pom-pom) is meant.

guns and manned by five officers and eighty-five rank and file. Each machine gun carried 1,350 rounds of ammunition and a reserve of 36,000 rounds was carried by the ammunition carts. At the beginning of the war the guns were mounted on carriages with high wheels of the general type of a field-gun carriage. Major Balck attributes to this method of mounting the exceptional heavy loss sustained by the Russian machine-gun batteries at the battle of the Yalu. The guns thus mounted made good targets for the Japanese artillery. Later on the Russians substituted a low tripod mounting with a shield, and this gave good results.

Major Barrett thus sums up the Russian experience of the machine gun in this war:

> Russian reports speak in glowing terms of their efficacy, observing that in some cases they were even better than artillery. The chief advantages claimed for the weapon are:
>
> 1. They can be rapidly placed in position inaccessible to artillery.
> 2. They are easily hidden.
> 3. They range quickly.
> 4. Their fire is overwhelming.
> 5. They can follow up a moving target.*

On the Japanese side batteries of six machine guns were attached to the cavalry brigade, and there were also batteries of six guns and sections of two with the infantry. A Japanese writer, Captain Takenouchi, says that machine guns were not freely employed in the actual fighting until after the experience of their value at the battle of the Sha-Ho. The incident here referred to is described very clearly by General Sir Ian Hamilton in his "Staff Officer's Scrap-book during the Russo-Japanese War," in an extract from his diary dated October 12, 1904. In the two volumes of this most interesting and valuable work there are hardly any references to the use of

* "Lessons of the Japanese War" (Q. Club Prize Essay, 1906), by Major Ashley W. Barrett, London Irish Rifles. *Journal of the R.U.S.I.*, July, 1906.

machine guns by the Japanese. Indeed, a reader of the book would imagine that this weapon played a very small part in the war. Probably Sir Ian Hamilton's silence on the subject is the result of the fact that at the time even the keenest officers in our own Army had little idea of the value of the new weapon. Sir Ian Hamilton is continually describing the effects of artillery and infantry fire, but he hardly notices what the machine guns were doing. The incident in the Sha-Ho battle was, however, so striking that he gives a detailed account of it, which we reproduce here:

On October 9 and 10, 1,500 Cossacks, with their battery of horse artillery, attempted nothing decisive, but hung about between Penchiho and Chaotao, as if waiting for the fall of the former place. On the 11th, Prince Kanin, with the 2nd Cavalry Brigade and six Hotchkiss machine guns, arrived at Chaotao, and thus anticipated the Cossacks in making a raid, which everyone here has consistently assumed they must make. . . . To-day (October 12) at 3 a.m. Prince Kanin marched on Penchiho. At Senkin Pass he had a skirmish, and drove the Cossacks back northward. As I have already noted, the Russians in their attack on Penchiho had been trying to envelop the place, and their extreme left had actually worked round along the River Tai-tsu, due south of the defence line. Thus on the extreme Japanese right the defenders were thrown back like the lower part of the letter **S** along the tops of the mountains, whose slopes ran down into the river, whilst the Russians, with their backs to the river and their faces to the north, were half-way up the slope, still endeavouring to effect a lodgment on the crest-line. After the skirmish on the Senkin Pass, the Cossacks fell back as far as the Tai-tsu-ho, where they still interposed between the advancing Japanese cavalry brigade and their own infantry, who on the northern bank were busily engaged with the defenders of Penchiho. On the nearer approach of Prince Kanin, however, the Cossacks shifted their position eastwards, still covering their unconscious infantry so far as to forbid the Japanese cavalry from making any attempt to cross the Tai-tsu-ho, but leaving it open to them to occupy some high ground on the southern bank which was within effective rifle range of the Russian camp on the other side of the river.

Prince Kanin is not the sort of man who would miss good chances, and certainly on this occasion he seems to have unhesitatingly seized the ripe gift offered him by fortune. Stealthily manœuvring his six machine guns into position on a high and broken spur which ran down to the water's edge, he suddenly opened a hellish rain of bullets upon two Russian battalions who, at half-past eleven o'clock, were comfortably eating their dinners. In less than one minute hundreds of these poor fellows were killed, and the rest were flying eastwards in wild disorder. Next moment the Maxims* were switched on to the Russian firing-line who, with their backs to the river and their attention concentrated on Penchiho, were fighting in trenches about half-way up the slope of the mountain. These, before they could realize what had happened, found themselves being pelted with bullets from the rear. No troops could stand such treatment for long, and in less than no time the two brigades of Russians which had formed the extreme left of Stakelberg's attack, were in full retreat. Altogether the six Maxims had accounted for, according to the first despatch, 1,000; according to the second, 1,300 Russians.

Machine guns had been found useful at an earlier date in the skirmishes between the opposing cavalry forces at the outset of the war. In the first engagements with the Cossacks the Japanese found that in the mounted combat against the heavier men and horses of the enemy, they were at a marked disadvantage. The Cossacks were also better trained in the use of the lance, and employed novel methods of fighting. General Akiyama tells how, in the first skirmish at Sho-ka-tou (May 30, 1904),† the Cossacks charged with the reins fastened to their belts and both hands free for the lance. They struck the Japanese horsemen on the rifle-arm with the butt, dismounted them, and speared them as they lay on the ground. The Japanese realized that their best chance was to have recourse to dismounted tactics, and beat off the enemy with rifle fire, supported by the machine guns. During the rest of the campaign the Japanese cavalry were almost entirely employed as mounted infantry. Nearly all the fighting was with the carbine, and the machine guns were always in the firing-line. The uniform worn by the cavalry was very like that of the infantry, and on more than one occasion its

* The guns were really Hotchkisses.

† *Rev. Milit. des Armées Étrangères,* May, 1908. (Lavau II., 605.)

vigorous dismounted attack led the Russians to believe that they had to deal with a considerable infantry force, when only the cavalry were engaged. Thus, at the battle of Wa-fan-gou (June 15, 1904), the decisive flanking attack was made by Akiyama's cavalry brigade—two regiments of cavalry and a battery of machine guns, supported by two field batteries. The Russians gave way under the impression they were being turned by a much more numerous force.*

In the general orders issued by General Oku, at the outset of the war,† he noted that ground once won must be held at all costs against the enemy's counter-attack, and that in doing this machine guns would be of the greatest assistance. On the other hand, if the advance was held up by the enemy's machine guns, an effort must be made to put them out of action by artillery fire. Oku suggests that mountain guns would have been particularly useful for this purpose. Probably his idea was that these small guns could be brought up under cover to close range to shell the machine-gun positions.

A writer in the *Militär Wochenblatt* (June, 1908), summing up the experiences of the Russo-Japanese War, points out the important results obtained by the use of machine guns:

> The machine guns were extraordinarily successful. In the defence of entrenchments especially they had a most telling effect on the assailants at the moment of the assault. But they also were of service to the attack, being extremely useful in sweeping the crest of the defenders' parapets; as a few men can advance under cover with these weapons during an engagement, it is possible to bring them up without much loss to a decisive point. The fire of six machine guns is equal to that of a battalion, and this is of enormous importance at the decisive moment and place.

One may perhaps doubt if six machine guns can give the fire effect of a battalion, but the statement is worth noting, as showing the impression that the experiences of the Russo-

* *Militär. Zeitung*, July, 1908. (Lavau II., 607.)

† *Ibid.*, August, 1908. (Lavau II., 608.)

Japanese War had upon German military opinion. We may probably see in this the reason for the considerable increase in the machine-gun equipment of the German Army in the years before the Great War. Several Japanese officers have published their impressions on machine-gun warfare, based on their experiences in the field. The gun used was the Hotchkiss. In more than one instance one finds statements that there were frequent interruptions of fire by the gun jamming. One writer goes so far as to say that continuous fire for any length of time could not be depended upon, and it was therefore all the more necessary to choose the target and the moment for opening fire with a good deal of deliberation, as one could only count upon a good effect if the fire told from the first moment. The most frequent breakdown was caused by the breaking of the extractor, but there were also frequent jams of cartridges in the feed gear. A Japanese report published in Streffleur's *Militärisches Zeitschrift* (May, 1909) and summarized in the *R.U.S.I. Journal* (February, 1910), gives interesting statistics of the mishaps to machine guns. At the battle of Hei-kou-tai those caused by breaking of extractors averaged one per machine gun, and jammed cartridges one for every 300 rounds. There was one case of a gun being disabled by a cartridge exploding while being passed into the barrel from the feed-gear. This is a very high average of failures, much greater, it would seem, than has occurred with any guns of the Maxim type, and one would imagine that the result would be a very poor fire effect, but the guns produced excellent results—a testimony to the thorough training of those who worked them. Of course broken extractors would be rapidly replaced and jammed cartridges cleared away. But the Japanese writer was so impressed with the danger of the gun being temporarily disabled at any moment that he gives it as his opinion that machine guns should never be used singly. "To do so," he says, "is dangerous, because a machine gun is placed either at the weakest or the most important part of a position, and such a point cannot be rendered safe by placing one machine gun there. There is always the danger that the gun may be put out of action by some failure of the mechanism, and this would have a serious effect on the general situation."

In the intense cold of the Manchurian winter it was found that the rapid rise of temperature, when the gun was fired, had a tendency to increase the chance of failure of the mechanism. The Japanese kept their guns covered with blankets between the intervals of firing. Night firing was frequently used. On the offensive, aim was taken by the use of marks set up during the daylight. But considerable difficulty was found in arranging the feed of the ammunition in the dark, and in a later stage of the war a small shaded electric lamp was used to light up the feed gear.

All the Japanese writers agree that the machine gun must be used in attack as well as defence. Questions of cover and ammunition supply make its use in defence much simpler and more effective. They recognize that in the infantry firing line, useful as the machine gun is, it is by no means easy to secure sufficient cover for the gun and the group of men serving it, and with its large *consumption of ammunition the* question of supply is a difficult problem. In Manchuria much of the fighting, however, was of the character of the entrenched battlefield, which has been the dominant feature of the great war in Western Europe. It was thus possible to push forward a number of guns and a large supply of ammunition to a new position during the night and dig them in under cover before daylight. There is a general agreement among the Japanese writers that the best results are obtained at moderate ranges.

As regards range, in order to develop the full effect of the fire of a machine gun, this should not usually exceed 1,100 yards. Firing at longer ranges reduces the effect and betrays one's strength and position prematurely to the enemy. In the defence it is advisable to let the enemy approach to within 220 yards, and then to open fire, especially when the defender is in a strong position. Beyond 1,100 yards the machine gun produces good effect on closed bodies of troops and on artillery on the move. In the battle of Te-li-ssu (Ya-fang-kou) good effect was produced upon dense columns of the enemy, even at 2,500 yards.*

The machine-gun batteries were provided with wheel transport, but on the bad roads of Manchuria it was found they could not keep up with the infantry, and pack-transport was gradually substituted.

Herr Ullrich, the correspondent of the *Kölnische Zeitung* with the Russian Army, gives some interesting instances of the successful employment of machine guns which he himself witnessed.† On January 28, 1905, near Lin-chin-pu, the Japanese attacked a Russian redoubt armed with two machine guns. A Japanese company—that is about 200 rifles—was thrown forward in skirmishing order. The Russians held their fire until the range was only 300 yards; the two machine guns were then brought into action. In less than two minutes they fired about a thousand cartridges, and the Japanese firing-line was literally swept away.

On February 27 the Russians attempted to surprise a railway bridge over the Sha-Ho. It was a bright night, and objects could be seen 500 yards away. The Japanese swept the bridge with four of their machine guns. The leading Russian company was almost completely destroyed, and the attack failed. On the same night the Japanese attacked a village on the river, bringing their machine guns into action

* *Journal of the R.U.S.I.* (February, 1910), p. 226. The ranges given in the original article in Streffleur's *Zeitschrift* and in the Japanese memoir would be in metres. The ranges named for action in attack 1,100 yards, and in defence 220, would be 1,000 metres and 200 metres. This is noted here, as otherwise the reader might wonder why it was suggested that the enemy should be allowed to approach to a range carefully defined as 220 yards.

† Herr Ullrich's article was considered sufficiently important to be reprinted in the semi-official *Jahrbuch* of the Germany Army and Navy.

at 1,500 yards. In this case, artillery might have been effective. The machine guns gave no result whatever, for the Russian garrison was perfectly sheltered among the houses. On the other hand, a few days later the Japanese attacked Lin-chin-pu and the adjacent villages, and made a sound tactical use of their machine guns. They stormed the villages, and then brought the machine guns up into them, with the result that they easily beat off repeated attempts of the Russians to recapture them.

Herr Ullrich, echoing the opinion of the Russian officers with whom he had spoken, says that the machine guns ought to rank as a separate arm, like the artillery.

"Whichever of the two opponents," he says, "has at his disposal the larger number of machine guns, has thereby at his command such a superiority of fire that he is able to give an effective support to his infantry, and that he can occupy a considerable front on the defensive, with smaller groups of infantry supported by the guns, and the infantry thus becomes more free to manœuvre and more mobile." They are most useful to protect the flank of a fighting line, and they thus economize the amount of infantry allotted to this task and so enable one to strengthen one's reserve. He notes that the Japanese use their machine guns very effectively as a support to the infantry in the attack, concentrating a heavy machine-gun fire on the decisive point at the critical moment. On the attack succeeding they rushed up the guns to secure the position or to pursue the retiring enemy with their fire. There were also cases when a Russian attack was apparently succeeding, but was driven back by a reserve of machine guns suddenly brought into action from the second line.

As to the use of machine guns in the pursuit of the retiring enemy after the capture of a position, he makes some important observations. Within the limits of the range of infantry fire he says that for this purpose the machine gun was more effective than artillery. This is on account of the difficulty of ranging with shrapnel against a moving object, and the superiority of the flat sweeping fire of the machine gun in comparison with the descending showers of shrapnel balls.

The commander of the machine guns must be in close touch with the infantry he is supporting, but he must himself have the utmost liberty of action. The guns must be so organized as to be as mobile as possible, and each group of two machine guns must be ready to act as an independent unit. The fight with the enemy's machine guns is entrusted to the artillery. It is not an easy business, as the machine-gun detachments can so easily take cover. The targets are difficult to find, and can easily change their position.

By the time that the battle of Mukden was fought the machine-gun armament of the Japanese Army had been considerably increased. In the First Japanese Army every infantry division had attached to it four batteries, each of six machine guns, on the tripod mounting. The personnel of the battery was made up of one officer and 74 men. The officer was usually a lieutenant, and the section commanders in the battery were sergeants, a sergeant-major acting as second-in-command. To each cavalry brigade a machine-gun battery was attached. One may say that after the battle of the Sha-Ho new batteries of machine guns were continually being sent to the front. A Japanese officer wrote that their very presence was an encouragement to the troops. They went forward constantly, he said, if they heard the rat-tat-tat of the Hotchkisses, and he even suggested that it was better not to bring them into action near the artillery, as then the men did not so plainly hear their reports.

It would be easy to multiply instances of the effective use of the Japanese machine guns, but one may sum up the experiences of the war by saying that there were few occasions on which they did not prove useful. A few of them were always attached to the advanced guard on account of the rapidity with which they could be brought into action in an encounter with the enemy's advanced troops. Captain Takenouchi lays it down as an accepted principle with the Japanese Army that just as a few guns from the artillery are attached to the advanced guard, so also it must have some machine guns. On the battlefield the guns were used not only on the defensive, but were also boldly pushed forward with the attacking infantry. Towards the end of the war there was the disposition to use the guns in masses, thus, in the last stage of the battle of Mukden no less than four batteries were employed to beat down the fire of a Russian detachment holding the buildings and enclosures of a Chinese farm. Under the storm of fire from the twenty-four guns, the Russians abandoned the position. The machine guns were also used with good effect in securing a captured position from a counter-attack. Japanese writers point out that it is easier to hurry up a reserve of machine guns for this purpose than to bring up a body of infantry which could deliver a correspondingly heavy fire.

One may say that with the experiences of the Russo-Japanese War the machine gun had at last come into its own, and obtained a recognized position as a necessary element in all armies. The significance of these Japanese experiences was not fully realized at once in all countries, but in Germany there was an immediate realization of the lessons of the Manchurian campaign, and it had hardly ended when there began the rapid expansion of the machine-gun armament of the German Army, of which we have seen some of the results in the present war.

In the years between the close of the Russo-Japanese War and the outbreak of the Great War in the summer of 1914, there was a remarkable output of machine-gun literature, especially in Germany and in France. The growing interest in the subject led to a keener study of the possibilities of the new weapon. There had been sufficient war experience to make the development of a sound tactical doctrine possible.

Much that was written at this time is still of permanent value. Something is to be learned even from the suggestions put forward in the earliest stage of the gradual evolution of a theory of machine-gun tactics. We therefore proceed to summarize the views of machine-gun specialists of various countries in the years before the Great War.

CHAPTER IV

THE EVOLUTION OF MACHINE-GUN TACTICS

In the preceding chapters something has been incidentally said of the tactics of the machine gun. It was some time before any military writers attempted to deal with the subject in a systematic manner. In the first years after the introduction of the new weapon there was naturally, from lack of experience of its use in war, considerable divergence of opinion as to its value, and as to how it could be best employed. The strangest ideas were put forward on the subject, even by practical men. There was, in some quarters, an unfortunate tendency to judge it by comparing the machine gun with the field-piece. It was not yet realized that the small-calibre machine gun delivers a fire *which is absolutely different from* that of artillery, and can best be described as concentrated infantry fire—a maximum of rifle fire with a minimum of front.

Perhaps those who persisted in comparing machine guns with artillery were influenced partly by the mistaken tactical use of the weapon by the French in the war of 1870, and partly by the fact that inventors were producing, besides the rifle-calibre machine gun, guns of larger calibre, the projectiles for which were small shells. Indeed, there was a time when inventors and manufacturers seemed to think that the greatest future lay before these shell-firing guns, and that the rifle-calibre gun might prove to be comparatively unimportant. There might, indeed, have been a greater development of the shell-firing machine gun, but for the fact that the invention of quick-firing artillery produced guns that could deliver a storm of heavy shells, giving a fire power with which that of the shell-firing machine guns could not for a moment be compared.

In the technical literature of this transition period, when the machine gun was not yet generally accepted as a necessary part of the armament of naval and military forces, there were, on one side of the controversy, enthusiasts who wrote as if the machine gun might almost supersede every other weapon; while, on the other side, there were obstinately conservative writers who seemed to regard it as a mere ingenious toy, that would be rather an encumbrance than a help to soldiers in the field. In a discussion at the Royal United Service Institution in these early days a distinguished cavalry officer went so far as to say that a machine gun would be a nuisance with a cavalry regiment, whose commander had already quite enough to think of in handling his squadrons in action. In the midst of this conflict of divergent and contradictory opinions the machine gun slowly won its way into general acceptance chiefly through two channels.

In the navies, in the absence of quick-firing artillery, which was still far in the future, the machine gun was introduced as a means of bringing a rapid fire to bear upon a hostile torpedo boat. In a torpedo attack made by a small boat of high speed, suddenly appearing out of the darkness at close range, the question of safety for the larger ship depended upon what could be done in two or three minutes, and a gun that during this brief interval could send out a stream of 1,200 to 1,800 bullets was naturally regarded as a godsend. With machine guns on board the ships of the Navy, it became obvious that they had another use. They were an effective weapon with which to equip landing parties, and thus the Navy obtained some experience of what the gun could do on shore. We may say that in our own service naval landing parties supplied most useful experimental material for the

Army, and thus cleared the way for the introduction of the gun into the sister service.

The other channel by which the machine gun won general recognition was its use in warfare against semi-barbarous tribes in various parts of the world. For the most part, the tactical methods of these wild races had not got beyond the stage of trying to mob an enemy with a rush of spears, and such tactics afforded to the civilized soldier the easiest of targets for machine-gun fire. We have seen that the experiences of the war of 1870 tended to discredit the machine gun. It was the result of incorrect reasoning from the data afforded by the war. In 1870 machine guns were a failure when it was attempted to make them a substitute for artillery, but they did useful work in the few instances in which they were used in an intelligent way. But these exceptional cases were disregarded, and for some time most soldiers held that, however effective they might be in savage warfare, machine guns could serve no useful purpose in the regular warfare of disciplined armies.

Russia was the first Power to introduce a large number of machine guns into its army. This was because at the time the Russian Army was engaged in a series of expeditions in Central Asia, in which its opponents were semi-barbarous tribes, whose methods of attack made the machine gun peculiarly useful against them. Switzerland also adopted the machine gun at an early date, and this was because in the defence of the country it was useful to have a weapon which could sweep with its concentrated fire mountain roads and paths to which the enemy's main advance would be necessarily limited. We have seen that it was not till after the Russo-Japanese War that it was generally recognized that the machine gun could be used in the warfare between regular armies, not merely on the defensive, but also in the attack, and not on exceptional occasions, but under conditions that would daily present themselves in a campaign.

During this long period when the machine gun was gradually winning its way into general acceptance, a body of useful experience was being collected, theory was tested by practice, and a tactical doctrine of the machine gun was being evolved. A literature of the subject was growing up, mostly in the form of papers and articles in technical and professional journals. The writers were either men who had actually handled machine guns in the field, or students of war who based their conclusions on the available records of the use of machine guns in action. In the earlier writings on the subject one naturally finds a number of unpractical suggestions, but, as time went on, these were gradually eliminated, and there is now something like a general agreement as to the main lines of machine-gun tactics. It is remarkable that some of the earliest writers on the subject were far in advance of their time, and had sufficient insight into the possibilities of the machine gun to lay down a theory of its use, which is nearly the same as that which is now embodied in the Service Regulations of European armies.

It will be of interest to trace the development of the tactical theory by quoting the opinions of some of the leading writers on the subject belonging to the various armies of the world. Some of these are men who have actually commanded machine guns in action; others are writers who have made a special study of the subject.

One of the most practical and useful studies of machine-gun tactics is the work of an American officer. Lieutenant John H. Parker, of the 13th U.S. Infantry, commanded a battery of machine guns—Gatlings—in General Shafter's army in the Cuban War of 1898. In the following year he published a very interesting work on the "Tactical Organization and Uses of Machine Guns in the Field." What he wrote was not based entirely upon his own experiences with the guns on active service, but he had also very carefully studied the then existing literature of the subject. He had formed a very high idea of the possibilities opened out by the new weapon. Thus he wrote:

> As demonstrated upon the battlefield, it is the most versatile arm of the service. At times it can take the place of each of the other arms, and it has, moreover, a distinct field of usefulness peculiarly its own. It is infinitely superior to artillery against personnel, forms the strongest possible infantry or cavalry reserve, is the backbone of advance and rear guards, and renders a position once captured impregnable against counter-assaults. It can operate offensively wherever infantry can penetrate, and can move as rapidly as the best cavalry. Its defensive uses are equally numerous and varied. There is no place about a fortification where it is not welcomed as a valuable auxiliary and reinforcement.

With remarkable foresight Lieutenant Parker predicts the coming developments of field artillery. He notes that in the fighting before Santiago it was found to be impossible to bring guns into action under long-range infantry fire. The work of the artillery, he said, would be done at increasingly long ranges, with elaborate scientific help towards securing accuracy of fire, and with brakes to absorb the recoil. It would be no longer necessary for the batteries frequently to change their position; the artillerist would become more and more "the cold-blooded man of science operating at long ranges where the optician aids the naked eye," often using indirect fire against an unseen target. The place of the light artillery that once pushed forward with the infantry would be taken by another arm—the machine gun, and this is his ideal of what the new arm ought to be:

> The machine-gun man must be hot-blooded and dashing. He must have all the verve and élan of the best light cavalry, all the resisting power of stolid and immovable infantry. He is not to reason on abstruse theorems, nor approximate difficult ranges; his part is to dash into the hell of musketry, the storm of battle, and to rule that storm by the superior rapidity and accuracy of his fire. The characteristics of the machine gun and its crew are therefore essentially different from those of the artillery of the future and, it may be added, from those of the artillery of the present. The line of reason-

ing that must actuate the conduct of the machine-gun man is essentially different from the artillery motif. The experience of the battlefield has demonstrated that a greater degree of independent action pertains to the machine guns than is the case with artillery.

The spirit of the two weapons is so different that the machine guns are hampered by being attached to artillery. Lieutenant Parker asks if the necessary alternative is to make them an integral part of the cavalry or infantry, and he argues that the best results will be obtained by giving them a position independent of both. In a cavalry or infantry unit the machine gun will be regarded as an auxiliary weapon entrusted for the time being to the officer who is to handle it in action. He may be an enthusiast for the weapon and make excellent use of it, but the chances are that it will be only a temporary duty, and he will afterwards pass on to other occupations: "Special aptitude, special talents, special characteristics are necessary, as well as special training, for the proper service of the machine gun." And the best results will be obtained if the machine guns are in the hands of men who have chosen this branch of military service as their special work, and permanently devote themselves to it. The conclusion is that the machine gun should be organized as a special arm of the service.

Lieutenant Parker differs from most other writers on the subject in suggesting that the machine gun should be organized in batteries or, as he prefers to call them, companies of three machine guns each. He lays it down that the tactical unit in machine-gun work is the single gun. It is this view that has led him to suggest a grouping of three guns in a battery. Most men who have actually handled machine guns take the view that the normal fighting unit should be two guns. This enables an uninterrupted fire to be maintained; if there is a hitch of any kind with one of the guns, it will generally be possible to keep the other in action while it is being set right. With two guns as the unit, if the battery organization is adopted, we have thus groups of four or six guns which can be subdivided into sections of two each. A three-gun battery is suggested by Lieutenant Parker on account of his exceptional view that the single gun is the fighting unit. He further suggests the grouping of the batteries in machine-gun battalions. These questions of organization have been touched upon here, as they are closely linked with the tactical views that he advocates.

Valuable as it is on the defensive, he regards the machine gun as essentially a weapon for the attack. Thus he writes:

> The offensive uses of the machine guns have been as fully demonstrated as those of any other arm of the service. They have been used at close quarters and against entrenchments; in the open, and under cover; with and without supporting troops; they have demonstrated their mobility and serviceability over any and every kind of ground, and in all kinds of weather. *Therefore, we state that on* the offensive they are useful in conjunction with every other arm, and may be used alone where no other arm could be so used. The offensive use will be treated with respect to infantry, cavalry, and artillery, and as an independent arm; with advanced guards, outposts, the cavalry screen, and in battle.

He suggests that with advanced guards the machine gun will supply a useful "stiffening" which will often make it possible to use a smaller number of men, or again by increasing the number of machine guns the advance guard may be given a striking force that will enable it "to take a more strongly aggressive tone towards the enemy." He even suggests, probably in view of the operations of a small force, that the machine-gun detachments alone may form the advanced guard. He believes that such a force, on coming in contact with the enemy, could certainly delay his movements and hold its position long enough to give time for the main body it covered to come into action. With cavalry, acting as an advanced guard, or on a raiding expedition, the place suggested for the guns is with the reserve or support of the vanguard. Here Lieutenant Parker has in view the support to be given by the guns to the mounted action of the force in case of an engagement:

It will generally be practicable (he says) to manœuvre the guns to a flank or the cavalry to a flank. A quick movement will unmask the guns, which will pour a terrific fire for a moment, and then will be masked by the charging cavalry. The guns will at once despatch forward to complete the marvel or to afford a rallying point under cover of which the cavalry can reinforce.

Under the head of outposts, Lieutenant Parker states that machine guns should always be used to stiffen the line, their place being with the supports or reserves on the selected line of resistance.

It adds greatly (he says) to the security of the main body and tends to remove a feeling of remoteness and helplessness, which is always felt by any body of men who are separated from the main body and placed nearer to the enemy for the performance of this difficult duty.

Guns on the outpost line should be placed so as to command the approaches and sweep with their fire selected areas of ground to the front, and should be kept ready for instant action, concealed as much as possible. He suggests that it may sometimes be useful to attach a machine gun to a strong patrol sent out to the front.

Before discussing in detail the action of machine guns in battle on the offensive, Lieutenat Parker thus sums up the principles that limit and direct this action:

Machine guns are expected to develop the fire action of good unshaken infantry, plentifully supplied with ammunition, always under perfect control and fire discipline, and utterly without nerves. They will also be expected to have all the mobility of the best cavalry at all stages of the battle. When they get into so tight a place that animals cannot live for pack or draft, the men will be expected to draw the pieces by hand for a sufficient distance to render the use of the animals again available. Usually this will not be necessary, except in seeking a more effective position to enter the fight. If the guns reach so hot a place as the one supposed, they will be in an effective position. It must be always remembered that if the enemy can reach us with his small-arm fire, we can reach him with our machine-gun fire equally as well. We can therefore make the machine guns effective wherever infantry fire could be used. Artillery fire is not to be considered. It is assumed that the machine guns will not be pitted against artillery at long ranges; it has been demonstrated that the machine gun has nothing to fear from artillery after it once gets within effective range. It can bring so effective a fire to bear that no men can live long enough to serve the pieces. But in the general case, the enemy's artillery is firing, or at least occupied by the artillery of the attacking force.

Lieutenant Parker is an enthusiast on his subject, and therefore contemplates the possibility of machine guns in favourable cases silencing artillery. But even he recognizes that this will not often happen. In connection with artillery, the duty he assigns to the machine-gun detachment is to serve as an escort to it in the opening phase of the battle. He holds that artillery in action with machine guns well placed on its flanks will be safe from attack, and proposes that when the artillery is pushed forward to a new position, the machine-gun escort should move first and take up its covering position on the new ground while the artillery is moving forward, thus making it safe against any possible surprise.

But all this is preliminary to the attack. *Once the infantry* is thrown into action, the *place of the machine guns is with it.* They will be used to economize the infantry by increasing the fire effect at the outset, and they will supply an effective long-range fire. In connection with this latter point, he quotes from Benson:

> There is a very important difference in the effective ranges of the rifle and the machine gun when the former is handled by large bodies of men which are very difficult to control in extended order in the excitement of battle. Machine guns can, from the control under which they can be held, open fire with effect at ranges where a very large proportion of the infantry ammunition would be wasted. . . .

In supporting the infantry attack, Lieutenant Parker suggests that the machine guns can be used freely from positions behind the firing-line, firing over the heads of the troops.

With properly trained gunners he holds that there will be no danger of hitting the advancing firing-line and there will not be the risk that arises when artillery is used in the same way, namely, the danger of short bursts of the shells, and further the disturbing effect of hearing the roar of the shells as they pass overhead without quite realizing where they come from. With machine guns, he says the men will generally not even know that their fire is passing overhead. But he does not limit machine guns to this use, for he recognizes that positions will be found for them in the firing-line itself. As to the actual firing of the guns he thus describes what he considers to be their ideal use:

> It must not be thought that these guns will keep up a continuous fire, that they will go on for an indefinite time pumping a solid stream of lead at one place, as some of the opponents of machine guns seem to think in their discussions of the amount of ammunition used by the guns. Such targets are but rarely presented, though they do show up for a few minutes at a time. When a target is presented, the machine gun whirls around and cuts in; as soon as its target disappears, it at once takes advantage of its superior mobility to catch up with the infantry and seek a new place to go in.

In the final stage of the attack the machine guns should be boldly pushed forward to close range in order to exert their utmost power in increasing the volume of fire poured upon the enemy. Their support at this moment and "the continuous rattle" of their fire has an inspiriting effect upon the attacking troops. If the assault is successful, they help to occupy the position and hold it against the counter-attack, or co-operate in the pursuit. In turning movements and flank attacks, the machine gun is particularly valuable, on account of the impression made by the large amount of fire it can suddenly deliver, and this with a minimum number of men engaged.

The orders to be given to the machine-gun men at the beginning of the battle need not be very explicit. The commander of the machine guns should know where the real point of attack is, and should be kept well informed of the progress of the action. With this information he should be able to get into action in an effective manner on his own account, and is likely to do better when not hampered by orders than when compelled to try to comply with instructions laid down probably by an officer who does not understand the true rôle of the machine gun in the battle. A very high grade of talent is demanded in order to be thus able to cut in alone, but a high grade of talent is demanded for the service of the machine gun at any time.

Like all men who have the true spirit of the machine gunner, Lieutenant Parker insists chiefly upon the use of its exceptional powers in the attack. As to its use on the defensive, he insists upon its fire action being supplemented by careful measures for the concealment of its position. This is easier to effect than in the attack. On the defensive there will generally be time to select positions which give a good field of fire and natural cover, improved by a little handwork. He insists upon the necessity of a regular organization of reliefs for the gunners, and a carefully arranged ammunition supply. Machine guns will form a valuable element in the reserve of the defence, and if the fight has to end in a retreat their fire action will be particularly useful with the rearguard. He concludes by once more urging that if machine guns are to be effectively employed, they must be in the hands of highly trained men and officers.

To ensure the full development of these ideas it will be necessary to call the best talent into the service of the new arm. The pioneers in all fields of human endeavour must be men specially fitted for what they are to do. Hundreds are competent to do what has been once done, but few are competent to originate. The machine-gun man must understand fully all the tactical uses of all the other arms of the service, and the exact relation and usefulness of his own in relation thereto. He must have the keen eye that takes in at a glance the salient points of a battle, and detects with unerring certainty the critical time and place. He must have the finest combination of nerve, patience, audacity and determination, and be prepared to fight always "to the last breath of horse and man." No part of the service demands higher qualities.*

One of the first countries to adopt machine guns was Switzerland. The Swiss Army is essentially a defensive force, and the positions it would have to hold are largely made up of difficult mountain country, where the enemy's approach would often be confined to the well-marked lines of the Alpine roads and the mountain paths zigzagging over exposed slopes, or running through narrow defiles. It is obvious that in such positions the machine gun would be an invaluable auxiliary for defence. In Northern Switzerland the country is more open, and here, in case of a violation of the frontier by French or German armies, the Swiss might have to undertake regular field operations. Their study of machine-gun tactics is therefore not confined to the mere consideration of the use of the gun in defensive mountain warfare.

In 1904 Captain Vuilleumier, of the Staff of the Swiss Army, published a very interesting study of the action of machine guns against infantry. He prefaced his essay by

* We have said that Lieutenant Parker is an enthusiast for the machine gun. The American Army Regulations for machine guns issued in 1908 do not adopt his conclusions. At the very outset the principle is laid down that the machine gun is intended only to give support to the other arms, and cannot be made the principal weapon in an action. Two thousand five hundred yards is laid down as the extreme limit at which machine-gun fire is efficacious, unless under the most exceptional circumstances. The maximum effect, however, is obtained at shorter ranges, about 1,200 yards. The rules laid down for the use of the guns in the attack are obviously based throughout on Lieutenant Parker's book. The regulations include elaborate rules for the guidance of umpires at manœuvres. It is pointed out that when they are engaged against infantry, the guns may possibly be put out of action by marksmen sent forward to pick off the gunners, and that it will therefore always be necessary for the guns to be protected against such an attack by an escort of a certain number of good shots armed with the rifle.

pointing out that in all the neighbouring countries the machine-gun armament of the armies was being rapidly developed, and if Switzerland were engaged in war she would have to take this new development into account. He notes that in the literature of the machine gun, up to the time when he wrote, there had been few attempts to work out in any thorough fashion the tactics to be employed either by the machine gunners or the infantry leaders opposed to them. In his study of the subject he takes largely the point of view of the infantry officer who finds himself opposed by the new weapon. But before going into the question of infantry tactics he discusses the capacities and the tactics of the machine gun.

There is no need, he says, to describe the various systems of mechanism employed in the different types of machine guns. It is sufficient to bear in mind that the essential characteristics of the weapon are its rapid fire, long range, accuracy in trained hands, and the fact that it requires few men to work it, and gives to this small group of men the power of producing an extremely high fire effect.

He takes its extreme range to be about 1,800 metres, or nearly 2,000 yards. At ranges shorter than 800 metres the sheaf of bullets from the gun is extremely concentrated, and the zone of effective fire is a narrow one. He considers that the most effective ranges for machine-gun fire are from 800 to 1,500 metres. Up to 800 metres the rifle fire of trained men may be considered to be thoroughly effective, but once this range is reached, only exceptional men are steady enough to make good shooting, and the high trajectory of the bullet makes the chance of error increase with every lengthening of the range. On the other hand, the machine gun fired from its tripod or carriage gives steady shooting even at long ranges. At 1,000 metres the fire of the machine gun is as effective as that of infantry rifles at 300.

Against certain targets the superiority of machine-gun

fire is particularly marked: for instance, against wide and deep targets, such as are presented by masses of men, or against a line in the open which can be swept from end to end by the machine gun pivoting on its support. The superior accuracy of its fire, especially at long ranges, has been already noted. There can easily be a misunderstanding about this. No one would assert that against a small target the machine gun would give more accurate shooting than a skilful marksman with the rifle, provided only a few shots were fired. But when it is a case of keeping up a prolonged and rapid fire the machine at once asserts its superiority over the man who becomes tired, nervous, or excited when he has to keep up a rapid fire for any length of time.

As to the question of the comparative fire effect of a single machine gun and a number of riflemen, he estimates that the fire effect of a machine gun is equivalent to that of from 100 to 150 rifles, and, as the range increases beyond 800 yards, the advantage of a machine gun will be still greater. He notes that with the organization then existing in Switzerland a Swiss machine-gun company of eight guns went into action with 78,080 cartridges, an infantry battalion with 130,560. Thus the machine-gun company was prepared to deliver a fire equal to that of about two-thirds of a battalion.

He further compares the fire of machine guns to that of artillery. There is first the question of ranges. At distances over 1,500 metres no comparison between the two arms is possible. At the longer ranges the machine gun, like the rifle, is helpless against artillery. The cannon has advantages which no small-calibre weapon can claim. The fact that artillery projectiles all act with explosive effect at long range naturally produces a considerable moral effect on the troops opposed to them, and with quick-firing guns the number of shrapnel balls discharged in a minute far exceeds that of the bullets that can be fired at the same time from a machine gun. A Maxim can be fired at the rate of 600 bullets in a minute. In the same time a quick-firing gun will fire twelve shells, which are equivalent to about 4,000 balls, each shrapnel shell sending a shower of bullets into the target, and scattering them over at least as wide a zone as that of the machine gun at long range. On the other hand, the advantages of the machine gun are that it offers a smaller target and can be more easily provided with cover. With a proper mounting for the gun the machine gunner can more rapidly change the direction and the range of the fire, the gun is more mobile, and it can be more easily transported under cover through broken ground, across hedges and ditches and through woods. In fact, it can go wherever the infantry can move, and can be brought into action at any point in half a minute. Finally, whether they are compared with infantry or artillery, the machine guns need a much smaller front in proportion to the fire they can deliver.

Captain Vuilleumier deals very briefly with the question of machine-gun tactics. He notes that up to the date at which he wrote very divergent opinions on the subject were to be found in the literature which it had called forth, so far mostly in the shape of pamphlets and articles in reviews. Some of this literature he sets aside as being inspired to a great extent by the advertising campaign of various machine-gun manufacturers, with a tendency to represent the machine gun as the most important of all weapons. But even among serious writers, there was considerable divergence of opinions. Thus he quotes the German Boguslawski as advocating the view that a machine gun would be of little use, except on the defensive, because in the attack, unless it could be entrenched, it was likely to present a marked target in the firing-line and be rapidly destroyed by the enemy's fire. On the other side, he quotes the French writer, De Montbrisson, who argues that it is in the attack that the special characteristics of the machine gun will secure the best effect of its fire. Artillery, says De Montbrisson, cannot be pushed forward at close range in support of the attack without risking extremely heavy loss, while the machine guns are light enough to be brought up under cover, keeping pace with the firing-line, and when the position is captured these will be at once available to strengthen it against the enemy's counter-attacks.

So far authors of the most divergent views on other subjects were generally agreed that the machine gun should be regarded as a supplementary arm to be attached to infantry or cavalry, and that it would be a mistake to organize independent units formed of machine guns and gunners. But there was a minority which held the contrary opinion, and predicted that before long batteries of machine guns would be as much a necessary part of an army as batteries of artillery. In support of this view he quotes Colonel Mariani of the Italian Army, who had made a special study of machine guns, and Lieutenant Parker of the United States Army, whose study of the subject we have already analyzed. Finally he notes that the new Regulations of the Swiss Army, lately published, lay it down that the machine guns will be organized in batteries, but that the commander of a body of troops can detach one or more guns to act with cavalry, cyclists, or small bodies of infantry who have been given a special task to perform, and he considers it likely that machine-gun batteries will soon form a part of the organization of other armies.

After the Russo-Japanese War there was considerable output of machine-gun literature in the military reviews of the Continent. Some of these articles were brief studies of special incidents in the war, but others were attempts to produce something like a methodical treatise on machine-gun work with the help of the new data afforded by the Manchurian campaign. In the spring of 1906 Commandant Guérin, of the French Colonial Army, published a series of articles in the *Revue des Troupes Coloniales*, under the title of "The Use of Machine Guns in the Armament of the Home and

Colonial Armies." Part of this essay was devoted to an historical summary and an account of the existing French machine-gun organization, with suggestions for its improvement. In this connection he was one of the first military writers to discuss in some detail the use of machine guns on armoured motor-cars. He enumerates among the advantages that a considerable number of horses would be dispensed with and made available for other services. There could also be a considerable reduction in the personnel. He suggested that one armoured motor-car would replace a group of twenty-five men, thirteen horses, and two waggons. There would be the gain of rapid movement, at least on the roads, and a better and more deliberate use of the weapon on account of the confidence given to the gunners by armoured protection. On the other hand, there would be certain disadvantages. The motor would be liable to the usual chances of a breakdown—chances which were more considerable in 1906 than they are with the improved engines of to-day—there would be stoppages due to tyre troubles, and the armoured protection could not be made very strong, as weight would have to be kept down. The armed motor-car would be a good target for shell fire, especially with high-explosive shells, and its size would make it difficult to find cover for it in the fighting front. It would be difficult to keep in touch with infantry moving over open ground, and it would probably be more useful with cavalry. His conclusion was that "it was not very likely that the machine-gun armed motor-car would supersede the machine gun mounted on a little tripod, which could be hidden behind even a bit of tall grass, and would not run the same risk of drawing the fire of the enemy's artillery." At the same time, he added that experiments with armoured cars should be persevered with, for there would be occasions in war when they might render useful service.

He begins his discussion of the general subject of

machine-gun tactics by noting the great diversities of opinion to be found so far among writers on the subject, some apparently thinking that the only real use of the gun would be on the defensive, others giving a prominent place in the attack. There was a further division between those who would like to see the machine guns used only as auxiliary weapons, attached to infantry and cavalry units, and the more enthusiastic advocates of the new weapon, who held that the machine guns ought to be formed into batteries instead. He remarks that the diversity in the existing machine-gun organization and regulations in the various armies gave proof of the existence of this diversity of opinion.

Commandant Guérin notes that in the Russo-Japanese War machine guns were used freely on both sides with marked effect. On the Russian side they were organized in groups of eight guns and in groups of six by the Japanese. The former usually attached them to brigades or divisions, the latter to individual regiments. They were employed both in the attack and on the defensive. In the attack their most notable use was in rapidly reinforcing the troops who had captured a position, so as to secure them against the counter-attack. Commandant Guérin seems to lean to the opinion that their chief use will be in the defence. He quotes the opinion of an artillery officer, Commandant Rouquerol, to the effect that in the attack there will be fewer opportunities for their effective intervention, and one must give up the idea of their finding a normal place in the infantry advance. As for the idea of their being brought into action behind the firing-line and firing over the heads of the men, the trajectory is too flat to admit of this being done. If they were placed in such a position, gaps would have to be left in the attacking line for the action of the machine gun; their fire would be limited to the narrow front thus defined, and the whole advance would be subject to a new and unnecessary complication.

Commandant Guérin is in favour of machine guns being worked in batteries. He cites in support of this opinion Lord Wolseley's criticism on the British manœuvres of 1898, and he also quotes the criticism of Sir Redvers Buller on the manœuvres of the following year, in which he said that there was a tendency to subordinate the tactics of machine guns too much to the action of the battalion to which they were attached, and that it would be much better if they were allowed to manœuvre more independently. In Commandant Guérin's view, the battery organization will give this necessary freedom of action to the machine guns. If they are organized in small regimental sections they will find themselves tied to a regiment or even a battalion. Grouped in a battery, on the other hand, they can either be used together or distributed in sections, according as the tactical position demands. And he goes on to say:

> The machine gun can render services of the highest value, but always on condition that it is used under such circumstances that its fire can develop its full efficacy. . . . *The use of the machine gun* on the battlefield cannot therefore be defined by one fixed rule. The machine gunner will often find himself in unexpected situations, and it depends on himself to select and take the necessary steps to meet the varying conditions presented by the course of the conflict. This applies to its action with all the other arms; a knowledge of the powers of the gun and the effects of its fire will be the guide for its co-operation with another arm, and officers must prepare themselves to use their weapon in battle, above all, by the study of various situations carried out as far as possible on the actual ground. No rules or regulations can replace this indispensable personal preparation on the part of the officers.

Machine guns should accompany the advanced guard and have an extra supply of ammunition with them. They would be of the utmost use to the advanced guard commander in enabling him to develop a heavy fire over a considerable front, and so assist in holding the ground until the main body can come into action. There should usually be a machine-gun officer on the staff of the divisional General or Brigadier. His business will be to see to the general direction of the machine-gun batteries or sections and provide for their ammunition supply during the battle. On the line of march during the advance the machine guns will be placed in a group, so that they can be readily moved out to chosen positions when the column of march deploys for action. In fact, their position on the line of march will be dictated by the same principle as that applied to the artillery. Once they are sent into action the utmost liberty of initiative must be given to the officers in immediate command of them.

When the deployment takes place, they will not, as a rule, be pushed at once into the first line. If they follow it up, it will be easier to select cover for the movement, and to choose a favourable position for opening fire.

During the prolonged struggle of the actual attack the machine gun may be able to play a very considerable part—supporting the assault on the position, helping to repel the enemy's counter-attack, or holding him by an intense fire when one is obliged to suspend the advance for a while and wait for reinforcements, or again using its fire to repulse counter-attacks after the position has been carried. If it is to fulfil these various tasks, the section of machine guns must be capable of extreme mobility and readiness to adapt itself to varying conditions. The officer in command must be a man of cool character with a good eye for ground, and ready decision in choosing the fire positions and the targets against which to use his gun, and in selecting the precise moment when he must change position and push forward as ground is gained.

Like all other writers on the subject, Commandant Guérin insists on the value of machine-gun fire in covering a retreat. He notes that the machine-gun commander before retiring should already have chosen another position on which to come into action, and take care that he is able to move on to it out of sight of the enemy, so that when he opens fire it will have the effect of surprise.

When acting on the defensive, it will be possible to make

elaborate arrangements for cover. The gun should be as completely concealed as possible; positions should be chosen commanding lines of approach, where the enemy will probably advance in force, and care should be taken not to open fire on targets that will give small results. Thus, for instance, it will often be a gain to keep the guns silent at the beginning of the hostile advance, whilst only the point of the enemy's advance guard or a thin line of skirmishers is in sight, and to wait until fire can be opened upon the main advance. Ranges should, of course, be measured beforehand, and if possible marked. In all cases the sudden and unexpected opening of heavy fire will produce more effect than the expenditure of the same amount of ammunition when the fire gradually develops and does not come as a surprise to the attacking force.

In night operations, machine-gun work will be difficult and exceptional. Commandant Guérin discusses the question of the use of shields for machine guns. In the war with Russia the Japanese used Hotchkiss guns, mounted on tripods, and fitted with a steel shield 70 millimetres (or about 24 inches) square, and 5 millimetres (about 3/16ths inch) thick, and weighing about 60 pounds. At short ranges it was found the Russian bullets pierced these shields. With the cavalry, shields were also mounted on the wheeled carriages of the machine guns. After the war there was considerable diversity of opinion as to the utility of the shields. It was recognized that they gave a certain amount of protection, but the shield on the tripod mounting made the gun less mobile. Commandant Guérin says that on the whole Japanese opinion seemed to be against the shield. He suggests the solution of the question might be summed up by saying that on the offensive the shield is more disadvantageous than useful, but on the defensive more useful than disadvantageous, and he proposes that the shield should be made in two parts carried by the transport of the machine-gun section, and fitted or not fitted on the gun according to circumstances.

He notes that towards the close of the war the Russians obtained a certain number of Madsen automatic rifles weighing about fifteen pounds, and fired from a rest with chargers containing from twenty to thirty cartridges. He gives no instances of their actual use in the war, but he suggests a weapon of this kind might be given to the infantry regiments in the proportion of two or four to each company. But he foresees difficulties in the supply of cartridges—difficulties provided for in the organization of a machine-gun section, but not so easy to avoid in case of automatic rifles placed here and there in the actual firing-line, and consuming a large quantity of ammunition. The soldier actually armed with the Madsen would have to be attended by a number of ammunition carriers, who would thus cease themselves to be combatants. He suggests, however, that the lighter weapon might sometimes replace the machine gun in colonial expeditions. He makes the suggestion that the automatic rifle might be part of the equipment of field batteries for their defence against a surprise attack at close quarters. He is opposed to the idea of attaching machine guns to the batteries as an escort. In this case, he says, the guns would be immobilized, and probably exposed to the same fire as the battery. They might, however, be attached to infantry detailed for escort duty with the artillery.

The accepted Russian views on machine-gun tactics, as formulated after the experiences of the war with Japan, are to be found in a summary of lectures given to the junior Military School in St. Petersburg, in 1906, by Lieutenant-Colonel Anisimov, of the Russian Staff. The Russian machine organization is described in detail on the basis of that employed in the Manchurian campaign—namely, batteries of eight guns, either mounted on wheeled carriages, or with tripods and pack transport. The author recognizes that the latter organization ensures a greater mobility.

Wherever a horse, or even a man, can go, the machine gun can go on its pack-saddle. In case of need it may even be taken off the horse and carried by hand. The machine-gun company with wheeled transport, on the contrary, can only move on ground which is practicable for waggons, and this considerably limits its field of action.

Colonel Anisimov describes at considerable length the special qualities of machine-gun fire. He lays it down that the machine gun is chiefly a weapon for medium ranges, and that at extreme ranges, in most cases, the results obtained do not justify the heavy expenditure of ammunition. But the ranges he names for effective fire are longer than one would expect from this preface. He says that experience shows that the machine gun can open fire "with complete success" against firing-lines in open order and their supports up to a range of 1,050 metres (over 1,100 yards); against the infantry lines in close order, and against cavalry, up to 1,400 metres; and against batteries, infantry columns, and cavalry in the open up to 2,800. He describes two methods of ranging, the one on the same principle as the artillery bracket; the other, by opening fire with a single gun at the shortest estimated range, and gradually increasing the elevation until the target is found. There were apparently no optical range-finders in use with the Russian Army in Manchuria.

As to the choice of position, the first point to be considered is a good field of fire, and then cover for the machine guns and their carriages. The guns should not fire over the heads of the infantry, but should be placed on the same front as the firing-line. Only in exceptional cases, where there is no need of a wide range of fire, should the guns be placed *en échelon* behind the flank of the infantry line, or in rear of an interval. Against long-range targets he allows that indirect fire can sometimes be employed.

The position should be reconnoitred in advance by the commanding officer of the machine guns. The guns should be brought up to a secondary position under cover in rear of it, where the guns should be taken from their limbers and brought up by hand to the fire position, to avoid prematurely attracting the attention of the enemy. The position of each gun should be chosen carefully and every effort made to hide it. Careful arrangements should be made for ammunition supply, and the waggons brought up under cover as near the guns as possible.

The general principle he lays down for the use of machine guns are: (1) Make the utmost use of cover by taking advantage of the ground and entrenching. (2) Do not open fire unless the situation justifies the expenditure of ammunition —that is to say, in general, at moderate ranges. At long ranges fire only upon important targets of considerable extent. (3) Make the guns as mobile as possible, so as to be able to act by surprise against the enemy; with the same object, conceal the presence of the guns up to the decisive moment. When fire is opened, let it be with as many guns as are available. (4) Have an infantry or cavalry escort for the machine guns, as they are not adapted for fighting at close quarters. (5) As far as possible the guns of a battery should be kept together. If this cannot be arranged, in no case should an isolated machine gun be brought into action, for a single gun may be temporarily disabled by a jam or a breakdown of its mechanism at the decisive moment.

On the offensive, machine guns can be used (1) in the reconnaissance stage, (2) in the period of the fire preparation for the assault, and (3) after the assault.

In the reconnaissance period a few guns can be sent forward with advantage to strengthen the fire power of advanced guards and reconnoitring detachments.

In the opening stages of the attack the machine guns should be kept together in reserve during the artillery preparation.* When the infantry advances the machine guns will go with it to support it by their fire. In open ground it is difficult for the machine guns to get nearer the enemy's position than a range of from 500 to 700 yards. The smaller is the target presented by the gun, and the lighter it is, the greater is its chance of being pushed up to close range.

From this point of view (says Colonel Anisimov) our machine guns, mounted on wheeled carriages, are quite unsuitable. Their high carriage, their shield and their heavy weight, all contribute to make them excellent targets for the enemy, and there is no doubt that during the advance to the attack they can be demolished at long range; consequently, in the attack the only practical type of machine gun is that mounted on the flat carriage.

In all cases, he says, an officer commanding a group of machine guns must have two ideas during this period of the attack: first, to push his guns up as near as possible to the enemy; secondly, if a chance offers, to get upon his flank and open an enfilading fire.

He should avoid becoming involved in a duel with the enemy's machine guns, but use his fire against important targets, the enemy's batteries, and reserves, and supports.

If the assault is successful, the machine guns should be rushed up to the captured position, in order to help to secure it against the counter-attack. If the assault fails, the machine guns will cover the retirement of their own troops and check the enemy's offensive, not hesitating to sacrifice themselves, if necessary, for this purpose.

In the defence of positions machine guns should be used to protect the flanks, to bring a cross-fire to bear upon the approaches to the front, and to augment the fire defence

* Colonel Anisimov has here in view a theory of the attack which is now becoming obsolete—namely, the idea that the battle should open with an artillery duel, and the infantry only come into action when on one or the other side a preponderance of artillery fire has been established, and the enemy's batteries are being silenced.

where the ground is most favourable to the attack. They are particularly useful if they can be placed so as to sweep a defile in the approaches to the position. Good positions can often be found at advanced points of the position where the guns, if well placed, can bring flanking fire to bear upon the attack. Their mobility enables them to be withdrawn in good time to the main position. They can also supply a most valuable reserve kept under the direct control of the commander of the whole position or one of its sections, ready to be rushed up to a threatened point. Such a reserve can also be usefully employed to extend the line and meet a turning or enveloping movement. It will be well on the defensive to divide the machine guns into two groups; one of these will be used for the general scheme of defence, and the other held in reserve to meet emergencies. Where a large number of guns is not available, it will be often well to use them all as a reserve.

Entrenched positions will be prepared for the guns, the men, and the ammunition waggons, so as to provide cover for them all, and for the movement of the ammunition carriers. Ranges will be measured and marked off, and if a telephone is available, it will be used to connect the machine-gun detachment with the commandant of the position. During the artillery preparation the machine guns will be silent, and their men will keep under cover. They will come into action as soon as the advance of the enemy's infantry presents a favourable target. But it is important not to keep up a uselessly prolonged fire, but to cease firing and wait to reopen fire again until it can be directed against a target likely to give effective results. Above all, a good supply of cartridges must be kept in hand for the final stage of the enemy's advance. It is then that the guns can render the greatest service. Their rapid fire at short range against the troops rushing to the assault is almost irresistible. On several occasions during the war with Japan, the assault was checked at the last moment by the fire of the machine guns which made it impossible for the enemy to push forward. This should be remembered, and the gun should be kept in action to the last moment firing at point-blank range.

If the assault is repulsed, the machine guns pursue the retiring enemy with their fire; if it succeeds, they assist in covering the retreat, and do what is possible to prevent the enemy entrenching himself on the conquered ground.

With cavalry, machine guns will in the dismounted combat be used in the same way as they are with infantry. In the mounted combat they will endeavour to take up a position on a flank from which they can assist the charge of the mounted men by firing upon the enemy up to the moment of contact. In broken country they will often be able to choose this flank position on ground where they are perfectly safe from the enemy's mounted men. Machine guns are sometimes assigned as an escort to artillery, but in this case they cannot entirely replace the usual infantry escort; in fact, they have need of it themselves.

Japanese opinion on machine-gun tactics is summed up in an essay by Captain Takenouchi, an infantry officer, who served with the machine gun in the war against Russia, and whose study of the subject was reproduced in the *Journal of the Royal United Service Institution* and the French *Revue d'Infanterie*, 1907. Captain Takenouchi is opposed to the wheeled mounting for the guns. It is practically impossible, he says, to bring a gun on a wheeled carriage into action. The gun must be fired from a tripod, and is best transported on a pack-saddle. The essential feature of the gun is that it can deliver a very heavy fire for a short space of time, and on a narrow front produce the same effect as a number of rifles that would occupy a considerable space. The gun is not fitted for prolonged action, and the machine-gun officer must therefore carefully select the occasions for its intervention. He considers that if one increases the number of guns in action together it does not follow that the aggregate effect of their fire is increased in the same degree. Two guns will often do all the work that is necessary, and there would be only a waste of ammunition in increasing the number to four or six. His idea is apparently based on the theory that the fire of the guns will usually be directed upon a narrow front, and the two guns will supply a sufficient intensity of fire. The section is thus the normal fighting unit. It is the minimum effective force, as a single gun is liable to be temporarily out of action, and in any case cannot fire continuously. He notes the special advantage of having machine guns with the advanced guard, which has so often rapidly to develop a heavy fire, but only a small number of guns are wanted in this position. Most of the guns should be with the main body. For instance, if there is a battery of six guns in a column, two should be with the advanced guard, and four with the main body.

On the defensive the guns are partly distributed on the main fighting-line, partly held in reserve. They will not open fire on the enemy in the first stage of the attack when their target will probably be only a thin firing-line, and they will inflict little loss and reveal their position to the enemy's gunners. They should reserve their fire until the final stage, when the enemy masses for the assault, and the sudden blasts of fire from the machine guns will inflict heavy loss and probably bring the attack to a standstill or completely break it up. In the attack the guns will push forward with the infantry and come into action when the occasion offers, especially when the neighbouring infantry ceases its own fire in order to rush forward to a new firing position. He remarks that on several occasions in Manchuria the sudden opening of fire by the machine guns produced a temporary cessation of the Russian fire, and enabled the Japanese infantry to push forward after having been for some time at a standstill.

One sees in these various opinions a general agreement

on the main points with divergences in matters of detail. We have seen that the results of the Japanese War, and the opinions expressed both by Russian and Japanese writers and by foreign spectators of the operations, led to the introduction of the machine-gun arm on a large scale in the German Army. In France the possibilities of the weapon were not so completely grasped, but the report of what was being done in Germany led to efforts being made to strengthen the machine-gun element, and at the same time a discussion of tactical questions began in the French military reviews and in a number of pamphlets by officers of the army. Probably the greatest impulse to the study of the subject was given by the remarkable work of Commandant Lavau, published in two volumes in 1908 and 1910. The title of the book is somewhat misleading, "Mitrailleuses de Cavalerie" ("Machine Guns with Cavalry"). Commandant Lavau, an officer of the 15th Dragoons, took up the study of machine guns, with special reference to his own arm. He saw in the gun the possibility of enormously increasing the fire power of a cavalry detachment, and he was not content merely to study the work of others, but made a long series of experiments as to the best method of mounting and transporting the gun with a machine-gun section of a cavalry regiment. It is evident that he originally intended his book to be a special treatise for cavalry officers, but in his enthusiasm for the subject he went far beyond his original scheme, and the two volumes are practically a collection of all that he could find on the subject of machine guns past and present, including such diverse materials as reprints of official reports on the one hand, and cuttings from current newspapers on the other. Throughout there is a running commentary from which one gathers his own ideas upon the subject. He believes in the machine gun as a weapon that will play an ever-increasing part in war; he hopes to see it recognized as a special arm of the service, and to see the *mitraillerie* taking its place, like the artillery, as an independent branch of the regular armies of the world. As there is a cavalry spirit and an artillery spirit which moulds the ideas and directs the actions of the cavalier and the gunner, so he says there should be a "machine-gun spirit" in those who handle the new weapon—a spirit of enterprise and dash, of readiness to seize the fleeting opportunities that offer themselves amid the confusion of the battlefield for the sudden intervention of the death-dealing weapon. For Commandant Lavau the defensive use of the machine gun is the least important. It is the weapon of the attack, and above all the weapon which will give a new strength to his own arm, the cavalry, enabling it to dispense with infantry support and even to some extent with that of horse artillery.

There is no doubt that Commandant Lavau's work has deeply influenced French military opinion. Besides the inspiration of his own enthusiasm, it gave to French military students of the subject an immense mass of material gathered from the most various sources, and embodying the experiences of officers of many armies who had handled or seen machine guns in action. In the years immediately preceding the war, other writers, sometimes drawing from this source, endeavoured to formulate a tactical theory.

Perhaps the best résumé of French military views as to the tactics of the machine gun is contained in a pamphlet by Lieutenant Dupeyré: "Nos Mitrailleuses, ce qu'elles sont, ce qu'il faut en attendre," published in 1912. The work professes to be, not an original essay, but a summary of the accepted views of machine gunners in France, in the period of preparation before the Great War. After describing the matériel and organization of a French machine-gun section, the author notes the special characteristics of the weapon. He describes machine-gun fire as having the advantages of great rapidity and precision, covering with its bullets a narrow zone of no great depth. The section of two guns is very mobile, easy to conceal, occupies a small front, and is not very vulnerable to hostile fire. On the other hand—

> The machine gun, being only a firearm, can only fight by its fire effect, and has not at its disposal shock effect, like the infantry and cavalry. It is thus an incomplete weapon, powerful no doubt towards the front, but weak towards the flanks, and as its very small personnel does not give it the means of guarding itself, and still less of defending itself at close quarters, the machine-gun section must always, as the official regulations lay it down, "act in the closest touch with the infantry of which it must be the auxiliary."

Following the French Regulations, Lieutenant Dupeyré notes that, as a general rule, the two guns of a section will fire alternately, thus securing time for the gun that is not in action to cool down, have its barrel cleaned out, and its mechanism oiled, if necessary. He says that experience with the French machine guns shows that after firing 1,000 rounds the barrel should be rapidly cleaned out.

On the march the guns should be well to the front of the column, so as to be able to come quickly into action when it deploys. The Regulations lay down that, as a rule, machine guns will be attached to advanced guards, flank-guards, and rearguards. They will be especially valuable with advanced guards on account of their power of developing a rapid fire, which will be useful in securing a position, and obliging the enemy to deploy, and checking his advance. Lieutenant Dupeyré quotes the French, German, Austrian, Spanish, and Japanese Regulations in support of his view that, as the preliminary action of the advanced guard draws to a close, and the main body deploys into line and begins the more serious engagement, it will generally be well to withdraw the machine-gun sections from the position which they have occupied to meet the first emergency, in order to assign them to the work they are to do in the actual battle.

In the first stage of the attack, in the zone of fire of the enemy's artillery, the commander of the machine-gun section must chiefly think of working under cover and relying on its invisibility to protect his section from the enemy's action. In this stage of the battle, however, good luck may give him a target for his guns, all the more because the battle front will not be an unbroken straight line, and, even at a very early stage of the advance, he may have the opportunity of bringing an oblique fire to bear upon the enemy in another section of the front, or he may catch sight of masses of troops hidden from the fire of those in their front, but open to such a cross-fire, or he may be able to enfilade an advanced section of the enemy's firing-line. It is for this reason that, even in the first stage of the attack, the machine-gun section must be ready to open fire at any moment.

Throughout the advance the machine-gun commander follows the movement of the troops to which he is attached, but he must be given the utmost liberty of choice as to when and where he will himself come into action. His little group of eleven men and two guns, forming the fighting front of his section, need not, however, at the outset align itself with the firing-line of the infantry. Once the range of 1,500 metres is reached, there will be opportunities of coming into action, but the machine gunner must remember that the Regulations lay it down that his guns are weapons for medium and short ranges, and one may conclude from this that they may delay to the last moment the actual movement into line, and endeavour to gain as much ground as possible in the advance without opening fire.

As medium ranges are reached, the opportunities for action will become frequent. Each must be seized as it presents itself. There will be no question of the guns firing continuously. This is, in fact, impossible. The support which the machine guns will give to the infantry will consist not of this impossible continuous fire, but of a series of short and violent bursts of fire against well-defined targets. Such bursts of fire can be used to support the forward movement of the neighbouring infantry, to raise its morale, and reinforce its fire when there is a check in the advance, to beat back a counter-attack of the enemy, to bring a sweeping fire to bear upon a strong point in the hostile line, forcing the enemy to keep down under cover and either cease fire or aim hurriedly and wildly.

As a rule the guns will not follow each forward rush of the neighbouring infantry. To do so would be to make useless changes of position, after each of which the range would have to be picked up again, while during the movement the guns would be out of action. If their position is well chosen, the machine guns need not move until the infantry has made perhaps four or five successive movements in advance. By this time the commander of the section will have chosen a new point to which his guns will be rushed forward as rapidly as possible. Lieutenant Dupeyré says that at manœuvres one sees machine-gun sections frequently moving not only to the front, but to right or left. These frequent movements would be impossible under fire, and would mean useless loss of men and waste of ammunition. Once they are in position, the guns must not be moved without good reason.

In common with so many other writers on the subject, he holds that it is when the range has shortened, in the final stage of the attack and before the assault, that the machine guns will do the most work. The stage of the fight has then arrived when there is abundant evidence to show that even well-disciplined troops fire hurriedly without aiming and generally fire high, so that their fire is practically harmless. It is then that the gun, fixed on its steady tripod, and working with the nerveless accuracy of a machine, pours its stream of bullets with unerring effect. At this moment the assault is imminent, and the guns "must open as heavy a fire as possible on the enemy's position, with the object of forcing him to keep down under cover in his trenches." The fire will cease as the charge rushes into the enemy's lines, and the machine gunners will then watch for any counter-attack on the flank of the assaulting troops, and when the assault succeeds, push forward to help in securing the captured position, and open a pursuing fire on the retiring enemy. If it fails, they will cover the retirement of their own troops.

With a rearguard covering a retreat the machine guns can render valuable services. Rearguard fighting is particularly well adapted to their power of suddenly opening a heavy fire, and the business of the machine-gun commander will be to choose, if possible, a position from which this fire will come as a surprise to the pursuing troops. Having accomplished his object of checking the enemy's movement and forcing him to deploy for the attack, he will fall back to another position where he can repeat the same manœuvre. Rearguard action is, as a rule, a fight on the defensive which is broken off at an early stage. In a prolonged defence of a position the machine-gun sections will be divided into two groups; those of the first group will be used to strengthen special points of the line. Here entrenched and carefully prepared fire positions will be made for the guns, the field of fire cleared, and ranges carefully measured and marked off. The rest of the guns will be kept in hand as a mobile reserve ready to reinforce any threatened point in the line.

Machine guns are helpless in frontal action against artillery in position, but they can be used with deadly effect against artillery on the move, and cases may occur where they can gain the flank of a battery in action, and in that case they will have the advantage of the artillery. They should not engage in a duel with the enemy's machine guns. Their normal target will be the enemy's infantry. But when engaged against infantry in broken ground the guns must have an escort of scouts, for instances have occurred where a machine-gun section has been put out of action by a few good shots gaining under cover a position from which they could open fire from a flank and pick off the machine gunners.

Machine guns in action are safe enough against cavalry attack, but they may be in serious danger if they are attacked while on the move. A machine-gun section, therefore, when it is not protected by the infantry with which it is acting, must have a few scouts to watch its flanks.

Machine guns can be usefully employed on an outpost line to sweep with their fire bridges, defiles, or other marked lines of the enemy's approach. They can do this even at night if marks have been set up in advance to direct their fire.

Lieutenant Dupeyré is opposed to the idea of employing machine guns as a support or escort to artillery. It is essential that the escort in such a case should be able to scout to the front and flanks of the battery, and it must therefore consist of infantry, but there are cases where a machine-gun section may be attached to the infantry escort of a battery.

Among English tactical writers one sees the same gradual approximation to a common doctrine on the value and use of the machine gun. At first English writers on the subject were chiefly interested in the discussion of the best way to employ the gun in our minor expeditions and little wars on the borderlands of the Empire. In these cases the problem was fairly simple. In fact, attention was chiefly concentrated upon the details of developing to the utmost the fire power of a few guns against the ideally easy targets supplied by wild rushes of semi-savage warriors anxious to come to close quarters at all costs. There was no need of going into the more difficult question of screening the guns from hostile artillery, or securing their effective co-operation with an advancing firing-line of infantry. The conditions under which we have to do most of our fighting almost dictated the system that was presently adopted of attaching one or two guns as auxiliary weapons to an infantry battalion or a cavalry regiment.

It was in the fighting on the North-West Frontier of India, where we had to attack enemies armed with good rifles and holding stone breastworks, that our machine gunners had first to consider the question of covering the infantry advance by using machine-gun fire to beat down the rifle fire of the hillmen. But even here the question was relatively simple, for it was generally easy to find a commanding position on the mountain ground which was the scene of the fight, from which the machine guns could be brought into action over the heads of the advancing infantry. The comparatively small use made of the machine guns in South Africa showed how unfamiliar most English soldiers were with the new problem of effectively using the gun in attacks in the open. Many practical soldiers were, in fact, disposed to leave quite out of consideration the problem with which so many Continental tacticians were busy—namely, that of the use of the machine gun in the attack against a regular European army, for until recent years many held that it was highly improbable that an English army would take part in European campaigns.

In the writings of the small group of enthusiasts who advocated the claims of the machine gun to be recognized as the coming weapon in the days before the South African War, one finds some remarkable forecasts of what would be its place on the battlefields of the future. But in the years after the South African War, and still more during the period of ever-increasing interest in machine-gun questions which followed the war between Russia and Japan, many of our soldiers devoted their attention to formulating a system of machine-gun tactics. Without attempting to give an exhaustive summary of this body of English opinion on the subject, it will be useful to note the opinions of some of these English tacticians. They were influenced partly by Continental opinion, but many of them brought to the study of the subject the practical knowledge of what the gun could do derived from their experience in our frontier wars.

General Hallam Parr, who had had considerable experience as a leader and trainer of mounted infantry, published in 1902 a little work on "The Future Training and Employment of Mounted Infantry and Yeomanry." Incidentally he dealt with the use of machine guns, and his ideas are so eminently practical that the whole passage may be quoted here:

In the attack the rôle of machine guns is an important one. They fill a gap between the most effective fire of the field gun, and the most effective fire of the rifle.

In practising the Attack, the sound tactical working of machine guns must be carefully watched. A good position, oblique to the general line of advance, should be, if possible, found at effective range from the enemy's position. There the machine guns and the escort must take up position as early as possible to assist the advance.

The great difficulty which always clogs the use of the machine gun firing S.A. ammunition at "extreme" ranges is, that it is so hard to discover whether the range has been obtained. To this the attention of the Officers in charge must always be directed, so that lead may not be pumped away into space. Sometimes a hint can be obtained from the artillery—sometimes a machine gun throwing a shell can assist—sometimes sandy or dusty patches will throw up some signs of the bullets striking. These aids, combined with the judicious use of a range-finder and telescope, are often all needed to throw light on the subject.

In coming into action the Officer in charge of the guns should try his utmost to bring them up unobserved. He should leave them well behind and get forward himself cautiously to look round and get the range, and choose exactly the positions. Arrangements to screen them should be madé *before* the guns are brought up, and he will consider it advisable if he is in touch with the enemy, to unhook out of sight and fire, and work his guns up by hand.

If he is in touch with the enemy he must act as if strong glasses were searching his position, and must act accordingly in order to deceive and confuse the enemy as to his intentions. If he is observed it may be as well to order the preparations for opening fire to be proceeded with, with perhaps some slight increase of bustle, and to go off himself to some other position, where he will be able to make his arrangements in peace, and to which he will be able to transfer his guns, when he is ready for them.

Without moving the guns too frequently, the Officer in command should seize every chance, as the fighting-line advances, of pushing his guns forward, and getting to close quarters to help the fighting-line, moving a certain proportion of his guns forward under cover of the fire of the remainder. Risk of losing guns should not deter him remaining in action when his fire is doing execution and rendering service. It is better to lose guns honourably than to save them when, by being in action a few minutes longer, success in another part of the field might have been assured or a retreat covered.

Directly the position where he is to come into action is reached, the Officer should see if he has not time and opportunity to improve it; ten minutes' spade work will do a great deal towards covering and hiding his guns and teams. He must not keep his escort close to the guns, or in too close a formation, and must remember to have patrols and scouts looking out for him on his flanks and rear. If the attack succeeds, the machine guns must reach the position almost simultaneously with the fighting-line, so as to make good the success.

When in a defensive position the Officer commanding the machine guns should take care to make his fire as effective as possible, morally as well as physically. He should make note of any open spaces the attacking force has to cross, and get some idea of the range by sighting shots with one gun, while perhaps his other guns are actually employed on the enemy. Sudden and violent resumption of fire partakes of the nature of sudden attack or surprise, and is most discouraging to the enemy, who has been congratulating himself on the lull which has taken place, and has probably imagined he has silenced his adversary's guns. Thus, before an enemy has reached a spot where the Commander of the machine guns intends to attack him, it may be well for him for a time to cease fire, and get his guns carefully trained and ready for opening a sudden and rapid fire.*

English military opinion was, however, still divided as to the real value of the machine gun, and therefore as to the method of employing it. We have proof of this in a lecture delivered before the Aldershot Military Society in 1904, by Lieutenant-Colonel W. D. Bird, then Chief Instructor at the Hythe School of Musketry.

Colonel Bird closes a brief summary of the history of machine guns by suggesting that the event which brought

* "Training and Employment of Mounted Infantry," by Major-General Hallam Parr, C.B., C.M.G., Aldershot, 1902, pp. 66-69.

home their importance to the military authorities of Continental armies was their adoption by Germany " in spite of the failure of the Maxim to produce any remarkable results in South Africa."

His description of the special qualities and defects of machine-gun fire suggests that he was more inclined to dwell upon its drawbacks than upon its favourable points. The lecture obviously belongs to a time when men who held a high official position in the Army were feeling their way on the subject. After noting the good points that the gun can deliver a great volume of fire in a very short time, and this on a narrow frontage, when it should be easy to conceal it, he notes that its mechanism is somewhat delicate and liable to be deranged at a most inconvenient time. It was liable to break down under the stress of prolonged rapid fire, about 10,000 rounds being the limit of consecutive action. It required a large quantity of ammunition, and the transport of this on a carriage or a string of pack-animals would detract from its mobility and give a conspicuous target. The water-jacket required constant refilling, and a peculiar noise of the gun in action easily attracted the enemy's attention. The beaten zone of fire was comparatively shallow, requiring very accurate range-taking. A relatively small error in estimating the range would greatly reduce the number of hits.

To neutralize the probable error in locating the target and taking the range, he suggests a means of sweeping an area of considerable depth and width by a method proposed by the German General von Rohne. The method is practically an attempt to spray a considerable surface with bullets by slightly traversing the gun during firing and at the same time continually raising and lowering the elevation. The suggestion belongs to the early experimental period, and better methods are now available. Range-finding is also a much simpler method than it was twelve years ago.

He holds that the machine gun cannot be compared with

the field or even the mountain gun: "the advantage rests so obviously with the latter weapons. It is merely a machine rifle, and must therefore be contrasted with the rifle." He quotes the German opinion that as regards mere rapidity of fire one Maxim gun is equal to about fifty rifles, but if other factors are considered to 100 or 120. Experiments at Hythe were taken to indicate that one Maxim was equal to about sixty rifles. This was the result of tests in rapid firing at 500 yards range, other tests at various ranges under field-firing conditions tended to show one gun was equal to twenty-five to thirty-five rifles, in the hands of good shots. On the whole, Colonel Bird thought that it was safest to value the Maxim in 1904 as equivalent to thirty rifles in the hands of first-class shots.

Certain points made in the lecture refer to methods then in use in our Army, but now obsolete. It was the time when Maxims were brought into action on wheeled carriages like field guns. This explains Colonel Bird's remarks that "the conspicuousness of the carriage necessitates the greatest care in bringing it into action"; and, again, the "weight of the carriage is against mobility, and the gun should therefore as a rule not be placed in the position when it may be required to move at a moment's notice."

He asks, Can decisive results be expected from the fire of thirty or even from sixty good shots? That is to say, from the fire of one or two Maxims? And he answers that "in ordinary circumstances it cannot." He concludes that "the Maxim gun is, in fact, a weapon not for general employment, but for use on special occasions when its timely intervention may be decisive."

He rightly points out that a Maxim gun should not be used singly. On the other hand, he is against the concentration of several guns in a battery, though he remarks this is the practice in Continental armies with the sole exception of Russia. If a number of guns are massed together, the in-

Brig.-General N. R. McMahon, D.S.O., p.s.c.

Author of "Fire Fighting."

Face page 95.

crease in fire power is outweighed by the increase in vulnerability. A large group of guns is certain to attract the attention of the enemy's artillery, and to be crushed by its superior fire power. They will be "exposed to annihilation." The whole argument points to the working of the guns in sections of two each.

One can see that in 1904 expert opinion at Hythe was still feeling its way in the matter of machine guns, but then came the new revelation of their power on the battlefields of Manchuria, and Lieutenant-Colonel N. R. McMahon, D.S.O., as the result of systematic experimental work, drew up the notes on machine-gun tactics that were embodied later on in the drill-books, and as we have already noted influenced German opinions and methods. Colonel McMahon had the merit of foreseeing the coming importance of the weapon. In a lecture delivered at Aldershot in December, 1907, on "Fire Fighting," he said:

> Even if automatic rifles are not adopted, machine guns will be used in the near future in very large numbers. There need be no fear of overstating the value of these weapons. All tendencies in modern tactics, night firing, envelopment, avoidance of open ground, cramped fire frontage, cavalry fire action, invisibility and mobile reserves, bring their good qualities more and more into relief.
>
> The Continent is no longer indifferent to the tactical qualities of machine guns. Russia has acquired several hundreds of Maxims, and Japan has issued regulations founded on Manchurian experience. Official Russian reports describe them as being of enormous value in close fighting. Their employment at close range would be rendered easier by the provision of small steel shields to protect the firers.
>
> A remarkable point made by a writer who served in Manchuria was the great falling off in the effect of rifle fire at short range, no harm being done at 150 metres or less, owing to nerves. It was at such ranges that machine guns were most useful with their nerveless carriages.

He pointed out that the objection formerly felt to intervals being left in a defensive line was disappearing, and noted that the borders of these intervals would be good positions in which to place machine guns to sweep with their cross-fire the ground in front of the opening. He forecasted the use of machine guns not merely in sections, attached to cavalry and infantry units, but as an independent arm. Thus he said:

> The war in Manchuria has shown that though as a rule machine guns are best employed in pairs attached to regiments, frequent occasions arise for using them as a reserve of fire power with advanced troops in pursuit or attached to a rearguard. The Japanese lay great stress on these functions of machine guns, which demand great mobility, and their guns will be organized, it is understood, in batteries, like those of Germany and Austria. It appears that machine guns, even those of infantry, should have great mobility and an organization favourable to massed action when necessary. They are invaluable, we are told, in advanced fire positions gained by night approach, and it seems likely that they will often be detached from regiments in our own service, as was the case in South Africa.

In Captain Applin's book on "Machine-Gun Tactics," written in 1909, we have a full recognition of the part to be played by the machine gun in the attack. It was now generally admitted that, valuable as it was for defensive action, this was only one aspect of its manifold possibilities. How far English military opinion has travelled in ten years may be seen in the reference to machine-gun tactics in an article by Captain J. F. C. Fuller, in the *Journal of the Royal United Service Institution*, November, 1914. Captain Fuller refers to the characteristics of the gun, "a nerveless weapon" controlled by one man, with an enormous volume of fire that is more accurate than that of the rifle, and can be diverted to right or left at will, while rifle fire is only to the front; and in describing the breaking of an enemy's line, he suggests that machine guns will be used to prepare the attack and cover its flanks against the counter-attack from right or left, which he regards as the best method for the defence to adopt.

> There is as much difference (he says) between machine-gun and infantry fire to-day as there was between light infantry and heavy

infantry fire a hundred years ago. So great is this difference that we might almost say that the light infantry of the future will be evolved from the machine gunners of the present. That is to say, that the assaulting column of the future will be flanked by these terror-spreading weapons, and that these new light infantrymen, like the old, will not only precede the assaulting column by working up close to the line of the holding attack, but will flank it on both sides, producing a somewhat similar effect on the hostile line as grape, canister, and case-shot did during the first fifty years of the last century.

CHAPTER V

MATÉRIEL

In a work like this, it is not necessary to go into the mechanical details of the various types of machine guns. These will be found fully described and illustrated in the handbooks for each variety of gun, either officially issued or produced by the manufacturers. One may say, indeed, that the mere details of mechanism are the least part of what the machine gunner has to learn. A thorough knowledge of the mechanism is of course necessary. It ought, indeed, to be so thorough that the gunner can take to pieces and assemble his gun, even when blindfolded, because he may have to handle it in the dark and deal by touch, not sight, with a temporary breakdown of the mechanism. But this knowledge of the gun from the mere mechanical point of view is the ABC of the subject. It is like the infantry soldier's knowledge of his rifle; when he has mastered it his real instruction as a rifleman begins. And the knowledge of the mechanism cannot be acquired from any amount of reading or examination of diagrams. It can only be learned by handling the gun, playing with it, pulling its action to pieces and putting it together again, firing it under various conditions, and gradually learning how to find out in a moment, almost instinctively, if anything is going wrong.

But, without attempting to describe fully and in detail the action of the various machine guns, it may be well to note and compare the leading principles and characteristics of the chief types of guns that are now in use.

Machine guns may be divided into two great classes: guns of which the mechanism is actuated by hand by means of a crank handle or lever; and guns the action of which is automatic. The guns of the former class are now almost obsolete, though it is quite possible that we shall again see machine guns in use, on improved lines, the action of which will not depend upon the automatic principle. For the time being, however, hand-worked guns are out of date.

All machine guns now in use in European armies belong to the second great class. Since Sir Hiram Maxim produced his first machine gun, inventors and constructors have endeavoured to reduce to a minimum the work actually done by the man who fires the gun. The modern machine gun is essentially an automatic weapon, and there is a tendency to extend the same principle to rifles and pistols, even with some of the very details to be found in previously constructed machine guns.

The automatic rifle may before long supersede the magazine rifle. It is, indeed, not always easy to draw the dividing-line between magazine rifles and the lighter patterns of machine guns. Thus, for instance, the light French Hotchkiss of the cavalry type, or the Lewis and Madsen gun, might be regarded as a repeating rifle when fired from the shoulder, and as a machine gun when fired from a tripod or other support. Perhaps the practical dividing-line may be taken to be that the machine gun is a weapon of rifle calibre adapted to be normally fired from a fixed support, such as a tripod, thus securing the uniform steadiness of fire that is impossible with a rifle fired from the shoulder.

Broadly speaking, all machine guns of the modern or automatic type may be divided into two classes:

1. Those in which the mechanism is actuated by the recoil of the barrel, of a breech block, or of a cartridge case.

2. Those in which, before the bullet has actually left the muzzle, a portion of the gas produced by the explosion is

made to act upon a piston by means of an outlet from the barrel.

It is obvious that in guns of this second class we have practically a small gas-engine attached to the gun, the explosion supplying the heated gas that works it. But it is equally true that in the guns of the first class we have also a gas-engine working. Thus, for instance, in the Maxim gun the recoiling barrel may be regarded as playing the part of a piston in transmitting the working pressure of the gas to the mechanism.

It is interesting to note that in the seventeenth century the celebrated French chemist, Papin, was engaged for a while upon a project for a gas-engine which was to be worked by the repeated explosions of small charges of gunpowder. But hearing of the work done in England by the Marquis of Worcester, with the steam-engine, he abandoned the idea, and turned his attention to elaborating a better type of piston than the Englishman had used. When Sir Hiram Maxim invented his gun, he unconsciously took up the line of research the French inventor had abandoned, and produced a gas-engine worked by gunpowder explosions, though this is not the usual description of his invention.

In all the guns of the second class, or of what we have called the gas-engine type, the piston is a single-action device, resembling in this the pistons of the early steam-engines—that is to say, the gas of the explosive acts only on one side of the piston and in one direction, driving the piston rearward. It is brought back to its original position by the action of a spring, which has been either extended or coiled up by the movement of the piston, and reacts as the driving force of the gas comes to an end.

In this respect these guns resemble those of the first or recoil class. In those also, as the recoil comes to an end, the action of the spring brings the recoiling portion of the gun back to its original position.

We have, thus, at each shot fired by a machine gun two movements in contrary directions: what may be called the backward movement, caused by the recoil or the stroke of the piston; and the return movement, resulting from the reaction of the spring.

This alternate backward and forward movement is simple enough. The range of movement is necessarily very short, for, if the action of the mechanism is to be rapid, the recoil of the barrel or the stroke of the gas piston has to be kept within a small space. In the Maxim the length of the recoil of the barrel is considerably less than the length of the cartridge. The problem for the inventor of machine guns is to supply a mechanism actuated by this reciprocal movement which will perform the cycle of operations involved in reloading and firing the gun. After each shot this will include the opening of the breech, the ejection of the fired cartridge, the insertion of a fresh cartridge, the firing of it, and so on, as long as the gun is kept in action.

The breech-opening device may be generally described as either the backward movement of a breech block by a lever connected with the piston rod in the gas-engine type, or in the recoil types the separation of the breech block from the barrel with guides or fixed cams to regulate its movement, and linked and jointed levers to accelerate it. The return movement, which pushes the cartridge into the chamber, cocks the firing action, and locks the breech mechanism, is invariably actuated by the force of a spring coiled up or contracted during the first movement.

The ejection of the fired cartridge has to be effected during the backward move. Many ejection devices are fairly simple, no more complicated, in fact, than those of the ordinary service rifle, and depending on a metal grip holding the base of the cartridge and jerking it out as the breech mechanism clears the barrel. In the Maxim and guns of a similar type the device is somewhat more complicated; for what is called the ejector has to do a double work, and not only take out the fired cartridge, but bring a fresh cartridge into position for reloading.

This is one of the characteristic points of the Maxim invention. The ejector is a narrow strip of steel placed vertically on the front of the lock, and free to move alternately upwards and downwards. On its front surface it has on each side a flange curving inwards, gauged to slide over the projecting rim at the base of cartridges of the usual pattern, or to engage in the groove at the base of a Mauser or Mannlicher cartridge. Cartridges can thus be held by their bases so as to project horizontally forward from the metal plate. At the outset, in preparing the gun for firing, the lock is moved backwards and forwards twice by means of the side-lever outside the gun. In this backward and forward movement curved levers on each side of the lock alternately push the ejector upwards and let it fall back into its normal position. On the upward movement, the flanges at the side of the ejector slip over the base of the cartridge, and as the lock swings backwards pull it out of the feeding belt. As the lock swings forward, the ejector drops, so as to bring the cartridge opposite the barrel into which it is pushed. On the second rise of the ejector, the cartridge is held in the barrel and cannot rise with it. But as the ejector rises, the flanges that hold the cartridge slide over it and grip another cartridge in the belt. The gun is now ready for firing, with one cartridge in the barrel, and another held ready to replace it. When the firing and the automatic backward and forward movement of the breech action begin, the same process goes on, cartridges being in succession taken from the belt, pushed into the barrel, and drawn out empty; and as the ejector rises to take a new cartridge, the empty one is pushed out of it into the ejector tube or allowed to fall directly to the ground. Thus the sliding mechanism on the front of the lock, though known as the ejector, is also the loader.

An examination of the Maxim action shows that the very short recoil of the barrel, which by itself would not leave room for the insertion of a new cartridge between the lock and the loading chamber, is made to actuate levers which carry the lock with the ejecting and reloading device sufficiently clear of the barrel—that is, to a distance a little greater than the entire length of the cartridge. This clearance is effected partly by the lock being made to pivot on a crank axis. In most guns there is some arrangement for thus giving the breech mechanism a longer movement than that of the recoil (or the piston stroke in the gas-engine type of gun). We have always varieties of the same device depending on the use of some kind of lever pivoted so that the power is applied to its short arm, and the work done by its long arm. This is, of course, a case of sacrificing power to secure speed of action. But, in the case of a machine gun, the force available as each shot is fired is far beyond anything that is required to actuate the mechanism. In fact, in most guns there are to be found stops and buffer springs placed to bring the backward action to an end. As this stoppage takes place, the spring coiled up during the first stage of the action begins to effect the forward movement.

Sir Hiram Maxim, in a letter addressed to Messrs. Loewe and Co., of Berlin (March 3, 1896), gives an interesting account of the various devices that can be used for actuating machine guns. He tells how, after his first series of experiments, he took out a large number of patents in 1884-85 covering as wide a range as possible. It was, in fact, an effort to secure the control of "every possible system of operating a gun by energy derived from the burning powder." His enumeration of the devices he elaborated may be quoted here as a useful summary of the subject:

Many forms of piston guns were patented, some having a piston directly under the barrel, and being supplied with gas from the bore of the gun. In some the gas was taken from the barrel near the projectile, in others it was taken from about the middle of the barrel, while in others it was taken from near the muzzle. In some cases the piston was co-axial with the barrel, and provided with a hole considerably larger than the projectile, the projectile passing through this hole, the gases expanding and moving the piston forward, developing the necessary energy for working the arm. In other cases the cylinder was firmly secured to the non-recoiling part of the gun and the piston attached to the barrel. The pressure pushed the piston back at the instant of firing, thereby generating sufficient energy to work the arm. A patent was also taken for working the gun by a vacuum produced by the escaping force of the gases from the barrel. Many patents were taken on different systems of guns having barrels which recoiled. In most of these the barrel and the breech block were locked together. Upon firing they both moved backwards through about an eighth of an inch, when a lever attached to the mechanism was brought into contact with a stationary point of resistance attached to the non-recoiling part of the gun. The lever, forming a part of the crank handle of the gun, when once set in motion, continued to move, performing a portion of the functions of loading and firing by the energy stored in the moving parts, at the same time compressing a spring which again gave off energy for completing the cycle of operations necessary for performing all the operations of loading and firing. In all the systems heretofore referred to, a portion of the energy was stored in the *vis inertiæ* of the parts, and a portion in the springs, which was afterwards given out to work the arm.

In all machine guns the force for bringing back the breech action to its normal condition is stored in a spring, and in most guns there are other springs in the mechanism. A spring is always a more or less undesirable feature in a mechanical device, though its use is often inevitable. A spring requires constant readjustment after it has been some time in use, it is peculiarly liable to break down, and, where it is exposed to heat and sudden variations of temperature, it is liable to lose its temper. It is claimed as a special merit for the Austrian Schwarzlose machine gun that there is only one spring in its mechanism. This is the large spring at the back of the gun which stores up the force of the recoil. The action of the gun looks, at first sight, rather complicated, but is really fairly simple. The barrel is fixed, and on firing the cartridge case produces a recoil of the breech block. A jointed lever on the top of the breech block is attached at the fore end to the case of the gun, and at the rear end to the breech block itself. A projecting spur on this lever drives back a plate which compresses the main spring. The movement of the breech block, the jointed lever, and the spring, effects the reloading, cocking, and firing, of the gun. The design was obviously suggested by that of the Maxim, of which it is a simplification. It has a water-jacket and a belt feed like that of the Maxim gun, and is generally classed as a variant of the Maxim device.

All machine guns have been lightened in recent years by substituting steel for gun metal in the casing and working parts.

Besides the broad division between recoil guns and gas piston guns, there is another line of division marked by the method of cooling adopted.

All designers of machine guns have a serious problem to solve in the question of the cooling of the gun. Heat is developed in the action of every machine, but in the case of the machine gun every shot fired means a high rise of temperature, and with rapid firing this is a cumulative effect which, if unchecked, would make the gun useless by raising the loading chamber to such a temperature that a cartridge would explode before the breech mechanism had time to close. There are other drawbacks from the overheating of the gun, even before this point is reached. There must therefore be some way of cooling it, and we thus have a further subdivision of machine guns into the water-cooled and the air-cooled class.

Most guns of the Maxim type are water-cooled by means of a metal jacket surrounding the greater part of the gun barrel kept full of water. The water-jacket system has the drawback that after a few hundred rounds have been fired the heat of the barrel sets the water boiling; unless the water is renewed, the jacket will soon be empty, and meanwhile a jet of white steam escaping from the safety valve may easily give away the gun's position. For a long time no effort was made to deal with this drawback of the water-jacket system. It seemed to be accepted as a necessary evil, though the remedy was fairly obvious. It has now been met by carrying the steam off to a flexible tube and passing it through cold water, so as to condense it. The condenser thus completes the water-cooling system and makes it thoroughly practical. It has the further advantage that with the condenser at work a large supply of water is not wanted. When the water-supply had to be continually renewed, it was often a serious question how to provide a sufficient quantity when fighting in arid semi-tropical country or on bare, dry mountain ridges. This was one of the reasons why the French War Office rejected the Maxim system and adopted the air-cooled Hotchkiss.

When the air-cooling system is adopted, the general principle, carried out in various ways, is to reduce the heating of the barrel by adding to it metallic flanges which act as radiators by exposing to the air a larger surface than that of the barrel itself.

The results appear to be less satisfactory than with the water-cooled gun. There is a certain gain, of course, in the matter of simplicity of construction, but the cooling is not so effective, and the water-cooled gun appears to be superior to the gun cooled by the mere addition of radiating flanges to the barrel. One has a practical admission of this fact in an essential feature of the design of typical air-cooled guns. In guns like the Hotchkiss and the Colt, the breech mechanism is so arranged that a "live round" cannot be actually placed in the chamber till the trigger is touched. A cartridge is then pushed home and fired by the same action. Thus, the gun never remains loaded with a live cartridge in the barrel ready to be fired, for there would be danger of its being prematurely exploded by the heated barrel.

Lieutenant Buttin, of the 97th French Infantry, though a strong advocate of the Hotchkiss system, is quite alive to the difficulties arising from the air-cooling arrangement.* He grants that with continuous firing the barrel is very rapidly put out of action; but he argues that such firing is never necessary in war, and that, if a barrel becomes overheated, it can be changed in less than a minute. In the French Army the guns working alternately give time for the one to cool down while the other is firing. But, from Lieutenant Buttin's statements of the results of experimental firing continued for not many minutes, it is evident that the Hotchkiss is liable to become very rapidly overheated in a way that seriously affects its efficiency, even if the fire can be continued.

He quotes an account of experiments made in Belgium, as described by Lieutenant Noel in an article in the *Revue de l'Armée Belge*:

> In continuous firing for five minutes the bullets between the third and fourth minute begin to fail to take the grooves of the rifling. And if the firing is continued to the seventh minute, all the bullets miss the rifling, and do not carry farther than 300 metres. The machine gun cannot fire effectively without interruption for more than three and a quarter minutes, and only up to this limit does it give regular groupings on the target.

This means that in less than three and a half minutes the Hotchkiss barrel, despite its air-cooling arrangement, has become so heated that its expansion widens the bore, so that the bullets do not take the rifling properly, with the result that its shooting becomes irregular and unreliable. And if the firing goes on for another three and a half minutes, the further expansion of the barrel makes it more and more difficult for the bullets to engage in the rifling. The firing becomes more and more inaccurate, until at last, about the end of the

* Lieutenant Buttin, "Notes sur l'Emploi des Mitrailleuses d'Infanterie dans le Combat offensif" (Paris, Chapelot, 1909), p. 61, etc.

seventh minute, the gun has for all practical purposes become a smooth-bore. The bore is so enlarged that the bullets slip through it without engaging the grooves of the rifling; they are simply blown out of it without any rotary motion, and with a velocity decreased by the explosion gas passing between them and the bore. The result is that they fall harmlessly a little more than 300 yards from the muzzle.

Lieutenant Buttin does not say what was the rate of fire; but supposing that it was at the rate of 500 shots a minute, the gun would be reduced to this inefficient state after firing 3,500 rounds. It appears that in the Belgian experiments barrels were prematurely worn out by continuous firing, an average of 6,947 shots producing this result. But it is pointed out that if the guns are not kept in action for more than two minutes at a time a barrel can fire at least 50,000 rounds without being made unserviceable. The destruction of the barrel in such a case as the firing of about 7,000 rounds continuously is the result of the wear and tear of the rifling under the grinding effect of the bullets, *while the metal of* the barrel is being made less and less resistant by overheating. One may grant that in action barrels would not be subjected to this strain. A change of barrel would be made long before the destructive point was reached. But the defect of a growing inaccuracy of fire, beginning after the gun has been about three minutes in action, is a much more serious matter. All the more serious when it is added that within seven minutes the gun becomes useless.

Besides changing the barrel, the French machine-gun regulations point out other ways of counteracting the overheating of the gun barrel It is laid down that water will not injure a heated barrel, and the gun may be dipped into it, or water may be poured through the bore. The caution is added that the water must be cleared out of the action, especially out of the gas piston, otherwise the gun will not work efficiently. It is curious to see these suggestions included in the regulations for a gun that is normally air-cooled, and which was adopted in order to avoid the water-cooling system; for we have here somewhat primitive methods of water-cooling suggested to supplement the air-cooling.

The Hotchkiss has apparently been used with satisfactory results by our French Allies during the war, because care has been taken to avoid any prolonged continuous firing from a single gun; but the drawback of the ordinary air-cooled type of gun is that this special precaution has to be taken, and this must tend to limit its power at critical moments, when a heavy fire has to be kept up at all costs.

But it would seem that these difficulties do not arise only with the Hotchkiss type. The Colt is another air-cooled gun, and experience in the war shows that it also has to be very carefully handled. After about 500 rounds have been fired the heat of the gun becomes intense. At first a light oil used to be employed for lubrication purposes, but it was found that with the overheating of the gun this either evaporated or was burnt out and carbonized. The result of this was that through the failure of proper lubrication the mechanism worked badly, and not only the rate of fire fell off, but serious strains tended to damage the working parts. The difficulty has been met by using a heavier oil and injecting small quantities into the gun mechanism at intervals in the firing.

But this only partly meets the difficulty. Continuous firing for a very short time tends to overheat the gun to a dangerous degree. The practice at present is to change the barrel, if possible, after about three belts have been fired. The barrel thus removed then cools down, and can be used to replace that which was in action. It is quite true that, with thoroughly trained gunners, the change of the barrel is a matter of a few seconds' work. But, all the same, the necessity of frequently changing the barrels, as the result of overheating, is a serious drawback. One can well imagine cases in which it would be all-important to keep up a heavy continuous fire from the guns, and in such a case the changing of the barrel would only be a resource of limited value, for the changes would be so frequent that there would not be time for the removed barrel to cool down to any appreciable extent. The guns would soon be out of action.

Nor are these the only difficulties caused by overheating. It is found that, when anything over 1,000 rounds is fired continuously, the springs are so affected by the heat as to become temporarily weakened, and if the firing is continued they may lose their temper and be permanently useless. Asbestos pads are now used to diminish the heating of the springs. The rate of fire has to be very carefully regulated, so as not to vary much from a rate of 450 shots a minute. If the fire is more rapid, the stresses on the mechanism become excessive, with the risk of a breakdown. With a slow rate of fire there is a difficulty of another kind. The breech mechanism does not act with sufficient speed, and as the barrel heats there is a risk of a cartridge being exploded while it is being pushed into the chamber of the barrel, before the breech-bolt is locked and secured. An explosion of this kind may do serious damage to the gun, and is also dangerous to the gunner.

That the Colt guns have generally given good results is due to the careful training of the gunners. But it is a drawback to have to handle a gun which is neither safe nor efficient unless a number of precautions are taken, and which, despite all precautions, can never give a continuous fire for more than a very brief space of time.

It is to be feared that, from these experiences with both the Hotchkiss and the Colt, one must conclude that, compared to the water-cooled gun, the ordinary air-cooled gun has a radical inferiority. What is meant by "ordinary air-cooled gun" is a gun in which the cooling of the barrel depends merely upon a few flanges or projections, acting as radiators, and increasing to a certain extent the surface of the barrel exposed to the air, without any arrangement to secure that a continual and ever-changing current of cool air is passing over these surfaces. If this current of air can be secured, air-cooling is effective. Thus, for instance, a Maxim gun can be used on an aeroplane without a water-jacket. The ordinary surface of the barrel is then sufficient for air-cooling, because the aeroplane and the gun are moving rapidly through cold air at a considerable height, and the breeze passing over the surface of the barrel effectually cools it.

There is a type of gun in which the air-cooling is effective, because it includes a special device for securing the continuous application of an air current to the cooling surfaces of the barrel. This is the Lewis gun. Its air-cooling arrangements are perhaps the most notable part of the whole design, and might be regarded as a separate invention from that of the gun mechanism, properly so-called, an invention that might, indeed, be applied to other air-cooled guns.

The gun is the invention of Colonel Lewis, who served for many years in the United States Coast Artillery. It is a gas-piston gun. The gas is taken from the barrel near the muzzle, and actuates a piston in a tube lying close beneath and parallel to it. The rearward movement of the piston coils a spiral spring by means of a rack on the piston rod and a toothed wheel on the circumference of the drum that holds the spring. This supplies the forward movement. The whole action is very solid and simple, and it can be dismounted with no other tool than the point of a bullet in a service cartridge. The gun weighs only 25¼ lbs. It is a light load for a man, and can be fired from the shoulder, but gives better results if the light support attached to it is used.

The air-cooling device of the Lewis gun may be popularly

described as an air-jacket with an arrangement for continually and rapidly changing the air inside of it, the heated air passing away at once, and cool air taking its place and keeping up the cooling process. It is made up of a casing of aluminium slipped over the barrel, and carrying seventeen plates radiating from it about 2 inches across and running the full length of the barrel. The whole is enclosed in a light steel outer casing. We have thus sixteen air-channels, triangular in shape, running the full length of the barrel, and bounded by surfaces of aluminium. The steel jacket projects beyond the muzzle of the gun, slightly diminishing in diameter, and on the muzzle an attachment is screwed, the shape of which directs the powder blast from each shot through the opening in front of the muzzle, drawing out the air from the radiator grooves and causing a rush of air to replace it through the openings at the rear end of the air-cooling device. The firing of the gun thus produces a continuous stream of air from rear to muzzle of the gun through the channels of the cooling arrangement. The aluminium flanges have not merely to distribute the heat to the air that drifts naturally around them, as in the Hotchkiss and Colt systems, but the cooling air is continuously fed on to them. The practical proof that the device works efficiently is to be found in the fact that water has never to be used on service to cool the barrel temporarily, and a spare barrel is not supplied or carried as part of the equipment of the gun.

In his book on "Aircraft in Warfare,"* Mr. Lanchester, discussing the value of the Lewis gun as an aeroplane weapon, carefully examines the question of the amount of heat developed by firing, and the quantity of air that must be supplied to the radiating surfaces in order sufficiently to

* "Aircraft in Warfare, the Dawn of the Fourth Arm." By F. W. Lanchester, Member of the Advisory Committee for Aeronautics, with an introductory preface by Major-General Sir David Henderson, K.C.B., Director-General of Military Aeronautics. London, 1916.

counteract the rise of temperature. He shows that in the Lewis gun the air current in the cooling jacket runs approximately at the rate of 100 feet per second, and that this passes a sufficient amount of air through the cooling jacket to keep the temperature within working limits. We have thus in the Lewis gun a novel device which is extremely simple, and which renders air-cooling practically as efficient as water-cooling, with the special advantage for aerial work that the weight is greatly diminished and the difficulty of water-supply disappears. In aerial work, however, cooling is a fairly easy matter. Maxims have been used in the air stripped of the water-jacket, for the very speed of the aeroplane supplies a sufficient stream of intensely cold air around the barrel.

Before leaving this subject of barrel-cooling, we may note an interesting development pointing to the possibility of an effective cooling system for machine guns different from both the water-cooling and the air-cooling systems. In 1898 a British inventor, Major Fitzgerald, devised and constructed a machine gun that gave remarkable results at its trials. The gun was hand-worked, and of the organ or battery type, having eight barrels fixed in two parallel rows—an upper and a lower—of four each. At a public trial of the Fitzgerald gun, the inventor fired 800 rounds in two minutes, and then opened the breech action and invited those present to put their fingers on the breech ends of the upper barrels. There was some natural hesitation about the experiment, a veteran admiral remarking that he was not such a fool as to burn his fingers for anyone's amusement. Our informant, however, ventured to put his hand into the breech action and touch the barrel ends. "It was a surprise," he says, "for the sensation of touching the metal was just like putting one's hand on a piece of cold marble."

Beyond saying that he used a chemical agency, Major Fitzgerald refused to explain his method, and a brass box at the back of the gun was always kept locked at the trials. He would not agree to patent the system, saying it would be better worked as a secret device. Business men, however, are shy of taking up unpatented secrets, so the Fitzgerald gun has remained in the long list of machine guns that have never gone beyond the experimental stage. But its history suggests that chemical cooling may yet be successfully adapted to the machine gun.

In most machine guns there is a general resemblance in the system by which the cartridges are fed into the loading mechanism. In the earlier machine guns, worked by cranks and levers, before the introduction of the Maxim system led to the general adoption of automatic mechanisms, cartridges were generally fed into the gun in a rough-and-ready way by being made to fall into the loading position by their own weight. The drawback of these loading arrangements was that stoppages sometimes resulted from the cartridge not falling accurately into line with the loading chamber. In the case of the Gatling gun the drawback was met by the invention of the American engineer Accles, which he called "the positive feed." The cartridges were placed in a drum, in which each was held separately in a slot, in a cylinder made to revolve by being geared on to the crank handle that fired the gun. The cartridge was thus kept parallel with the axis of the gun at the moment when it was released from the feed, and passed into guides which carried it into the loading chamber.* This was the best feeding arrangement of any of the old type of guns, but it could not be applied to the Maxim, the inventor of which devised the much simpler and more effective arrangement of feeding the gun by means of a moving belt.

* The Accles "positive feed," not depending on gravity, could be worked from the side of a Gatling gun as well as from the top. Gatlings with the feed working from the side were made for the caponnières of the new Danish forts at Copenhagen. The guns had to be mounted in a confined space under a low roof, and there would not have been room for a top feed.

Most guns now use a fabric belt feed. In the Maxim belt the cartridges are held in position by being slipped into a series of pockets formed by clipping together two thicknesses of canvas or similar material. In the early days belts were made to hold 333 cartridges for some strange reason, but for over twenty years those for Land Service hold 250 rounds. The official description is as follows:

This belt is formed by two pieces of webbing connected together by eyelets and brass strips of two lengths, the projecting strips showing how far the cartridges should be inserted; the belt is made thick at the edge next the bullets by being folded over a piece of cord, so that the cartridges may be kept parallel in passing through the feed block, and lie evenly in the ammunition belt boxes.

The fabric belt for the Colt has the pockets for 250 cartridges woven in the webbing itself, and the only metal parts are two brass tongues, one at each end, to facilitate entry into the feed block. The Schwarzlose gun has a fabric belt feed, and probably is made to hold 250 rounds. The feed block is so made that the belt can be disengaged instantly, after any number of cartridges have been fired and without having to pull it all through the feed block.

Metallic belts, containing 250 cartridges each, are used by the Hotchkiss, Bergman, and Vickers guns. The belts for the above three systems of guns are so much alike in construction that the official description of that for the Vickers gun can be taken to cover the other two:

Metallic belts also are provided. These consist of 250 steel plates, the sides of which are turned up to form two clips, the rearmost one of which grips the cartridge case and the front one the bullet. The cartridge is pressed into the clips from the rear, and when the front end of the neck of the case is in contact with the rear end of the front clip the cartridge is in the proper position. The plates are joined together by brass split pins which make a flexible joint. At either end of the belt is a tag. Short lengths of belt, holding twenty-five cartridges, also are supplied. These can either be used as belts, or units can be detached from them to replace damaged units in the 250-round belts.

Keep the belts dry if possible; should they get wet, lay fabric belts out to dry and dry metallic belts with a cloth, subsequently wiping the latter over with a slightly oiled rag. New or stiff fabric belts should be well plugged.

The *metallic charger band*, containing either twenty-five or thirty cartridges, is the principal system of feed used in the French Services. (Three systems of piston guns which use the above are used in France; Hotchkiss, Puteaux, and Modèle 1907, varying only in minor details.) It is a flattish band of brass, with two rows of perforations; the metal from these stands up at right angles to the band in such a manner as to hold all the cartridges as requisite. These charger bands can be refilled a great number of times in the field by a charger filling machine. In the infantry equipment, twelve of these charger bands, of twenty-five cartridges, are carried in a box, six of such boxes (300 rounds) are carried on a pack horse (1,800 rounds).

Wet fabric belts have a tendency to swell, thus making both the work of the filling machine and the feed block very bad, if not impossible. Some years ago it was claimed that a special solution had been found which waterproofed the fabric belt. But it was not taken up officially in the British Service.

Some of the lighter guns, which approximate to the type of the automatic rifle, do not use the belt feed, but carry their cartridges in magazines of a more compact type. Thus, for instance, in the Lewis gun the magazine is a flattened drum which drops on to a pivot above the breech action, gears on to it, and revolves as the gun is fired, bringing each of its 47 cartridges in succession into a position in which the loading mechanism grips and pushes them into the chamber of the barrel. It is an invention quite independent of the old Accles "positive feed," and is much more compact and effective. But it is based on the same general idea. In the case of all these magazine devices, weight and size are a bar to any considerable number of cartridges being included in a single magazine; but it is claimed that the change of magazines can be so quickly effected that a high rate of fire can be maintained. The same plea is also put forward for the magazine that some French writers urge in favour of the short metallic charger bands of the Hotchkiss gun—namely, that it is easier to regulate the rate of fire and check expenditure of ammunition than with the long belt carrying hundreds of cartridges. But there is certainly a gain in being able, if only occasionally and in emergencies, to fire several hundred shots in succession from a single belt without having a second gunner watching the fire of the gun and supplying cartridges in order to maintain it.

The Italian Perino gun—a variant on the Maxim system—has metal charger bands, holding twelve rounds each. It is said that there is a device for feeding these from a hopper, with an arrangement by which each band as it is fed into the gun automatically links up with the next, and draws it forward till it in turn links on to the next.*

Machine guns now range up to 3,000 yards, and, though indirect fire is mostly used at these long ranges, the fact that the range of the weapon has so increased makes the question of telescopic sights a practical one. They have not yet been introduced in our army, but in the German Army they have been long in use. The sight is a short Zeiss prismatic telescope. When detached from the gun it can be used for reconnoitring and observing fire effect.

* Particulars of the loading will be found in Perino's British patent, No. 6949, of 1907. Charger bands of cartridges are placed one above another in a box attached to the left side of the breech action of the gun. In the device, as figured in the patent specification, there are five of these metallic charger bands, each holding twelve cartridges. There are special arrangements to regulate the descent of the upper bands of cartridges as the lowest band is exhausted. But the patent does not show any device for linking the bands with each other. Each band, as it reaches the bottom of the box, is pushed by hand into the breech action of the gun.

The mounting of a machine gun is of importance only second to that of the gun itself. A special study of British machine-gun mountings by Major Longstaff is appended to this chapter. Here it is only necessary to make a few general remarks on the subject.

The tendency to regard machine guns as a new kind of artillery led to their being mounted for many years on a carriage like that of a field-piece. This kind of carriage was singularly unfitted for the new kind of gun. At first sight it seemed to give it mobility, but in action it tended only to hamper it; for it made it a heavy load, diminished the chance of finding good cover, whether on the move or in position, and restricted its field of fire. The machine gun only became a useful weapon for general service when the gun-carriage arrangement was abandoned, and it was decided that the gun must be fired from some low mounting, like a tripod. The tripod was first introduced as a machine-gun mounting in connection with machine-gun sections attached to cavalry or organized for mountain warfare. Pack-saddle transport had been recognized as the best method of conveying machine guns under these circumstances, and the tripod was designed as a portable mounting which could be carried with them on the pack animals.

Once this method was introduced and the advantage of firing the gun from a tripod was fully realized, it was inevitable that this kind of mounting should come into general use. The old field-gun type of carriage having been thus rendered obsolete, it became a question of how the guns of infantry machine-gun sections were to be carried on the line of march. Pack-saddle transport, though it has the advantage of being able to go anywhere where a man can walk, has the drawback that a much smaller aggregate load is carried by a given number of pack-horses than the load that the same number could carry if it were conveyed in wheeled vehicles. There was a gain in using wheeled transport, especially in countries where there is an abundance of good roads. After various experiments, most armies* adopted the system of carrying machine guns, their tripods, ammunition, and stores, on waggons or carriages, from which they were taken on going into action. Thus, in our own army, a typical pattern of limbered waggon carries two machine guns with their tripods and a first supply of cartridges.

A tripod mounting has the great advantage that it can be easily adjusted, even on the roughest ground. In the German Army the mount used for the machine gun is four-legged, the four supports being jointed on to the frame, so that the gun can be placed at varying heights from the ground. When the supports are lowered to their lowest position, they form handles by which the gun and its mounting can be carried by two men, like a stretcher; but it is generally dragged along the ground, and is designed for this purpose, whence its popular name of the "sled-mounting." A somewhat similar sled, but without the adjusting joints, is used for dragging ammunition along the ground. On the line of march the gun and its mounting is fixed on top of a limber waggon which conveys the ammunition.

* With exception of France, Austria, and Switzerland, which keep to the pack-horse.

OFFICIAL BRITISH MOUNTINGS FOR MACHINE GUNS

By Major Longstaff

The following official publications have been consulted, in addition to the *Journal of the Royal United Service Institution:*

List of Changes in War Material. Published monthly in Army Orders.

[Note.—There is only one series of numbers in the above list, and it has already reached five figures. For example, § 3322 refers to No. 3322 of List of Changes, and concerns introduction of ·45-inch Gatling gun carriage and Gatling limber into the Service. The entry will probably proceed to say, "and drawings have been sealed to govern manufacture."]

Treatise on Military Carriages and Other Manufactures of the Royal Carriage Department. Fourth edition, corrected up to January 1, 1888; 515 pp., Index.

Treatise on Military Carriages and Stores connected with them. Sixth edition, corrected up to April, 1902; 648 pp., Index; separate volume of plates.

Hand-Book of the ·45-inch Five-Barrel Nordenfelt Guns, Marks I. and II. 1888. By command of the Lords Commissioners of the Admiralty. 23 pp.

Hand-Book of the ·45-inch Maxim Gun. 1894. By command of the Lords Commissioners of the Admiralty. 32 pp., including 12 plates.

Gunnery Drill-Book for Her Majesty's Fleet. Official, 1889. 180 pp. Nordenfelt and Gardner Rifle Calibre guns are dealt with on pp. 96-115.

Hand-Book for the ·303-inch and ·303-inch Converted Maxim Machine Guns (Magazine Rifle Chamber). Mounted on Carriages, M.G., Infantry; Carriages, Parapet; Tripod and Cone Mountings. 1911. War Office. 118 pp., 40 plates. (Reprinted with Amendments, 1915.)

Priced Vocabulary of Stores used in His Majesty's Service. Land Service Stores and Stores common to Land and Naval Services. Part II. 1909. War Office. 730 pp., Index.

All the dates, unless otherwise designated, will refer to those in List of Changes.

This short historical review of Official British Mountings for Machine Guns is only possible on the assumption that the reader has access to the *Hand-Book for the ·303-inch Maxim Machine Gun*, 1915, as above.

The Gatling gun (·45-inch) trail and limber, Mark I. (§ 3322), was under the field artillery in all respects. In the "Manual of Field Artillery Exercises," 1877, official, pp. 380-387 give detail for "Drill with Gatling Gun," manœuvring of horsed limbers just like field artillery. In the year 1883 Captain Lord C. Beresford, R.N., said there were 142 Gatlings in the British Service. The wheels for the above gun were 4 feet 8 inches in diameter, while the limber was described as of similar construction to the 9- and 16-pounder limber, Mark II. It was fitted for single or double draught, with an off and near shaft of the ordinary description for field limbers, but of lighter construction.

There was a lighter form of trail and limber for Naval Service Gatling and Gardner guns, the wheels being 3 feet 6 inches in diameter. It was fitted with a pole with two slats for man draught.

WEIGHTS (EMPTY) OF NAVAL AND MILITARY CARRIAGES.

Naval gun and trail	(about)	4	cwts.
Naval limber	,,	4	,,
Military gun and trail	,,	5	,,
Military limber	,,	8	,,

It was at this stage of evolution that Captain R. H. Armit (late Lieutenant R.N.), 22nd Middlesex Rifle Volunteers, came on the scene with the Alt magazine carriage, self-contained on two wheels with no separate limber.

On January 22, 1886, Captain Armit said: "The Central London Rangers (22nd Middlesex R.V.) was the first infantry battalion to use the machine gun in any form. At first our guns were mounted like field-pieces, with trail and limber; but we soon found that in close infantry formations this way of mounting rifle-calibre machine guns rendered them useless. For, whenever the guns had to be brought into action, they first had to be reversed and the limber taken to the rear, the ground covered by the operation rendering it impossible to work them in close formations. . . . Colonel Alt (commanding 22nd Middlesex) was thus enabled to arrive at the conclusion that rifle-calibre machine guns would have to be worked from a limber or magazine carriage; and having explained to me his views, I, in conjunction with Mr. T. Nordenfelt, from whom we had obtained our guns, designed in 1884, for Colonel Alt, the carriage we now have in use. The Alt machine-gun carriage is a combination of limber and trail on two wheels. The limber forms a magazine, protected in front by means of a Cammell compound armour plate, ½-inch in thickness, through which no rifle bullet (·45-inch lead) can penetrate. The limber or magazine carries 5,000 rounds, and on either side are racks for the rifles of the gun detachment. The trail is supported by four men when on the march, the drag ropes being hooked on to the axles, as in the naval field-piece. The gun is mounted on the limber between the wheels, and can be trained through an arc of 180 degrees without moving the trail. Gun, carriage, and ammunition only weigh 10 cwts. Having obtained the guns (and magazine carriage), the next thing to do was to train men to work them; and Colonel Alt having placed the guns under my command, I proceeded to compile a drill from our old naval field-piece drill."

Captain Lord C. Beresford, R.N., said on January 22, 1886: "Captain Wilson, R.N., who gained the V.C. in the Soudan, devised a method of mounting the gun (Gardner) on two wheels in very much the same manner as that shown by Colonel Alt. When we went up the Nile the other day, we asked Captain Wilson's permission to mount our two guns (Gardners) in the same way. We left the limbers at Cairo; we took our guns on the two wheels. There was a light pole across the trail which 3 and 4 carried, and when we were near the enemy the muzzle was always to the front; that is to say, at the word 'Action' 3 and 4 dropped the trail; the gun was immediately in action, and the feed was on the gun with the cartridges in it. . . . Some years ago Mr. T. Nordenfelt assisted me to mount the gun (Nordenfelt) on a galloping carriage, so that the gun would train over the wheels perfectly clear—in other words, so that it was always in action, and I compared that gun (and mounting) very much to what a torpedo-boat is to us at sea."

Captain R. H. Armit replied to the above: "I must mention that Captain Wilson, V.C., inspected our guns (Nordenfelt) at headquarters. He inspected Colonel Alt's carriage with the present (1886) Director of Naval Ordnance after El Teb and Tamaai, and designed his two-wheeled magazine carriage after having seen the former. I was present when the Director of Naval Ordnance visited our headquarters. I had the men with the carriages before him, and that was long before any such carriages were introduced into the navy, and Admiral Hopkins states that it was entirely to Colonel Alt that we owed the introduction of this carriage into the Service."

The first official pattern of *Naval* Carriage after that of Colonel Alt was § 5302, sealed on July 20, 1887, for Nordenfelt or Gardner. "The body of the carriage consists of a wrought-iron frame, formed to support the two steel ammunition drawers, and a wooden under box which is fitted with a steel cover. The frame is partly covered with a steel sloping top or shield. . . . At the rear end of the body is fitted the wire net receptacle, which is suitable for carrying light stores such as drag ropes. . . . The wheels are 4 feet in diameter, with 2-inch tyres. This field carriage is adapted to be drawn by horse or by hand as required, and is provided with the means of carrying 1,200 rounds ready placed in twenty-four Nordenfelt hoppers or Gardner holders. No limber is supplied for carrying extra ammunition, as this can always be more conveniently done by means of pack animals or country carts according to the circumstances of the locality."

The first official pattern of *Military* Carriage after that of Colonel Alt was § 5653, sealed on November 30, 1888: "Carriage, Field, Machine Gun, Infantry, Mark I." In § 5653 it was stated, "A small number of these carriages have been provided for service. Wheels 3 feet 9 inches in diameter, track 3 feet 11¼ inches. This was similar to § 5302, but longer in body and with heavier wheels. There was also a naval mark so like § 5653 that it is difficult to tell them apart. Note that all the above three carriages were only for Nordenfelt and Gardner guns.

When the ·45-inch Maxim was introduced, the carriage § 5653 was adapted for it, by removing the steel sloping top, and bolting an early form of the cone mounting on the resulting flat surface. This conversion is shown in the Naval Maxim Gun (·45-inch) Hand-Book of 1894. This carriage had an iron prop stick under the netting to steady the platform in action.

The first *Military* Carriage designed for the *Maxim* with Infantry was § 7142, June 26, 1889: "Carriage, Field, Machine Gun, Infantry Maxim, Mark I." See Plate XV. in Maxim Book, 1911. The wheels were 4 feet in diameter, and the gun was 3 feet 3 inches from the ground when in action, so there was not a field of fire of 180 degrees without moving carriage. The prop sticks were under the shafts. The Mark II. (Plate XVI., 1911), § 9079, was sealed on June 16, 1897. The wheels were 4 feet 8 inches in diameter, the gun 3 feet 6 inches from ground, and there were seats, etc., for two men. This mark was a distinct retrograde step on the Mark I. It was too heavy, the arc of fire was more restricted, and the wheels were of immense size.

The Mark III., § 10,647, February 6, 1900, had the same wheels, etc., as Mark II., but all the sixteen belt-boxes were on the same level, eight on each side of gun, and there were no seats on the carriage. This was the mounting upon which the author was first trained.

NOTE.—If places had been provided on this carriage for a Mark IV. tripod and the spare part box, a good practicable field equipment would have been the result. A few of the most imaginative machine-gun officers in South Africa in 1900 did make this change, and, what is more, used the heavy wheeled carriage as the dummy to draw the Boers' fire, while the Maxim was used on an improvised tripod away to a flank.

The Mark I., Waggon, Limbered, General Service, § 13,702, June 19, 1906, marks another stage in the evolution of the rifle-calibre machine-gun carriage. It is the first form of carriage which is not a mounting; it is simply transport for the guns and matériel. The author considers the infantry carriage should so carry the gun that it can be in action either on or off the same.

An experimental galloping carriage for the Nordenfelt machine gun was mentioned by Captain Lord C. Beresford, R.N., on January 22, 1886.

The first official pattern of Cavalry Carriage was § 6088, sealed on April 30, 1889: "Carriage, Field, Nordenfelt Machine Gun, Cavalry or Mounted Infantry, Mark I." It consists of a two-wheeled carriage fitted with shafts, and is drawn by two horses, one of which is ridden. It has seats for the two men who serve the gun. A pair of prop sticks are hung to the splinter-bar. The wheels are fitted with brakes. The ammunition-box holds 1,998 rounds in seventy-four paper-feed boxes.

The seats may be turned up during action to form shields. The carriage is connected to the axle-tree by two waggon springs.

The first Cavalry *Maxim* Carriage was § 7141, April 8, 1889: "Carriage, Field, Machine Gun, Cavalry or Mounted Infantry, ·303-inch Maxim, Mark I., also ·45-inch Maxim." It is constructed to carry either of the above guns and fourteen belt-boxes. It is arranged to give elevating, traversing, and oscillating motion to the gun, the last to compensate for difference in level of the wheels. The body of the carriage is built of steel, and consists of a frame to which is attached coach-springs, splinter-bar, shafts, framework, with front plate to which is fixed the top of carriage. A seat is fixed over the splinter-bar, on which the number firing the gun may sit, his feet passing through an opening in the frame and resting on a footboard underneath.

Mark II. carriage has the carriage springs removed, the body being supported over the axle-tree on rigid steel stays (§ 10,053, sealed on July 12, 1898).

The Mark III. reverts to the trail with limber type—that is to say, four wheels instead of two per gun (§ 10,924, sealed on December 22, 1900). The wheels are 4 feet 8 inches, with a track of 5 feet 2 inches. There are eight belt-boxes on the trail, four on each side of the gun, with a wire network in the usual place. In action No. 1 sits on a cycle saddle on the trail. The gun has the infantry pattern shield. The limber carries eight belt-boxes, and has a pole for two-horse draught, one ridden.

The first official Tripod was § 9608, sealed on December 20, 1897: "Mounting, Tripod, Maxim, Mark I." It says: "This mounting, of which a number have been made for special service, will take either a ·303-inch or ·45-inch Maxim gun, and is intended to be carried by pack transport." It has a traverse of 28 degrees. For illustrations see Plate XXIII. in 1915 Maxim Hand-Book.

The second Tripod was § 10,070, sealed on January 30, 1900. See Plate XXIV. in 1911 Maxim Hand-Book: "Mounting, Tripod, ·303-inch Maxim Gun, Mark II. For pack transport." This was much used in South Africa in 1900. The tripod was not adjustable in the same way as the Mark IV. is. But it is evident that the Mark IV. was evolved from the Mark II., and from no other.

The Mark III. Tripod, § 11,050, was sealed on October 26, 1901. It is practically the same as one of the mountings in British Patent No. 16,081, A.D. 1891, taken out by F. E. D. Acland, late Captain R.A. This patent includes designs similar to Carriage, Field, or Tripod, M.G., Maxim, ·303-inch, Pattern 1, Plate XVIII. of 1915 Hand-Book; also Pattern 3, Plate XX. of same book. The cycle wheel mounting also in the patent was not adopted for the army. An extract from the complete specification is given:

The object of my said invention is to provide a light and strong tripod mounting, conveniently transportable upon wheels, or upon horseback, or upon the back of a man.

According to one part of this invention, the gun mounting is constructed with tripod legs, the front legs being hinged or pivoted to the rear or trailing leg, and so arranged that they may be folded up and strapped to the said rear or trailing leg when travelling. The trailing leg is preferably made telescopic to facilitate the carrying of the mounting on horseback or on the back of a man as aforesaid.

According to another part of this invention, a gun mounting such as that above described is provided with wheels to facilitate its transport. Light wheels, such as those termed "cycle wheels," may be used, and in this case the said wheels are arranged so that they may, if desired, be raised from the ground when the front tripod legs are placed in the firing position, to relieve the wheels from stress during firing. When travelling, the front legs are folded and strapped to the trail as above described, the mounting then resting on the wheels.

Some of the patterns included by Patent No. 16,081, A.D. 1891, were in use by the Swiss machine-gun mounted companies *circa* 1905; as well as in the trade catalogue of the Deutschen Waffen- und Munitionsfabriken of Berlin at the same time, which supplied many of the smaller armies of the world.

The present tripod, Mark IV., § 13,336, was sealed on January 23, 1906; for illustration see Plates XXVI. and XXVII. of 1915 Hand-Book. One of the most necessary additions is a horizontal dial-plate, to be on the top of the socket and with an adjustable zero, so that, once the four tripods of a section are placed on the ground, each zero can be fixed on the same magnetic bearing. There should also be an adjustable stop on the dial on each side of the crosshead arm, in order to limit traversing fire, etc.

The first form of Parapet Carriage, Mark I., § 5307, was sealed on July 12, 1887, and it was much like the Mark II. on Plate XXVIII. in 1915 Hand-Book. The Mark I. was designed to take the Gardner or the Nordenfelt guns. It said: "The carriage is arranged to combine a steady adjustable rest, from which the gun may be fired over a parapet, with a wheeled carriage on which the gun may be readily moved as required."

The first form of Parapet Carriage for the *Maxim* Gun was Mark I., § 6215, sealed about 1888, and the Mark II., § 7705, was sealed in 1895. This will still be found in our forts. A few special carriages were made for Gibraltar parapets, and these were Mark III., § 10,881, sealed on December 30, 1899. For illustrations see Plates XXX., XXXI., and XXXII. in 1915 Hand-Book.

The Cone Mounting, Mark I., § 7369, was sealed in 1894, about which time it was made available for land in addition to naval service. See Plate XXXIII. in 1915 Hand-Book.

Carriage, *Field, or Tripod*, M.G., Maxim, ·303-inch, Patterns 1-3, § 12,724, sealed April 13, 1904, to March 23, 1905; see Plate XVIII., etc., of 1915 Hand-Book. These had a limber for one horse in shafts. A certain number of these were issued to volunteer units.

It should be borne in mind that from 1897, at least, pack transport was provided for the tripod mountings for use in hilly country. But little use was made of the pack saddlery at home for training of sections.

CHAPTER VI

MACHINE GUNS IN THE BRITISH ARMY

WE have already said something of the introduction of machine guns into the British Service in preceding chapters on the evolution of the machine gun and of machine-gun tactics. Here we shall only note a few further points which will serve to complete the record.

About the time of the Crimean War inventors were busy suggesting new warlike appliances to our War Office, and amongst these proposals various designs for machine guns were submitted. They were all of the mediæval organ-gun type, with gun-barrels placed side by side on a frame horizontally. They were very heavy and clumsy, and none of them were adopted.

In 1860 Colonel Martin submitted a design for a machine gun to the War Office. Two years later Mr. Palmer and General Vandenburgh, both of them Americans, proposed another invention of the same kind; and in 1866 the War Office considered a third type of gun, the joint invention of Mr. Dupuis and Captain Warlow, R.A. None of these inventions were considered to be of practical value.

The first gun that the War Office regarded as of any serious importance was the Gatling. In 1867 the Ordnance Select Committee tested it and tried it against a nine-pounder field-piece, with very fair results. But no steps were taken as yet to introduce it into the Service. In the same year our Military Attaché at Paris forwarded a machine gun designed by M. Mouceux. It was an organ gun, with twenty-one barrels arranged in three tiers of seven each—a clumsy contrivance which was rejected by the War Office. In 1868 Colonel Claxton's "rifle battery" was tried—a machine gun with eight, ten, or twenty barrels to be fired in pairs. It did not give sufficiently satisfactory results. In 1869 Major Fosbery, on behalf of the India Office, investigated the claims of the Montigny type of machine gun manufactured in Belgium, and made a favourable report upon it.*

The adoption of the mitrailleuse by the French Imperial Army in 1870 called general attention to the question of machine guns, and the War Office appointed a Special Committee to investigate the matter. They made extensive trials of the best of the machine guns then available, and presented two reports—the first in November, 1870, the second in November, 1871.

In the first of these reports the Committee expressed a decided preference for the Gatling type of gun, and recommended that a small number of them should be purchased. Twelve Gatling guns of the ·45-inch calibre for land service, and twelve of the same calibre and twenty-four larger guns of ·65-inch calibre for the Navy, were ordered as a tentative measure until further experience would be gained.

When this first report was made very little information was available as to the practical results obtained by machine guns in the Franco-German War. Before the second report was issued in November, 1871, the Committee had not only carried out further experiments, and obtained from our foreign legations information as to the intentions of Continental

* These details of inventions submitted to the War Office, and the account which follows of the proceedings of the Special Committee appointed in 1870, are based upon information contained in a pamphlet by Captain (afterwards Colonel) J. F. Owen, R.A., "Compound Guns, Many-Barrelled Rifle Batteries, Machine Guns or Mitrailleurs" (London, 1874). Captain Owen was then Captain-Instructor at the Royal Gun Factory in Woolwich, and took part in the proceedings of the Special Committee.

Governments with regard to machine guns, but had also taken the evidence of a number of officers who had been with the French and German armies during the war.

Colonel Fletcher, who summed up this last kind of evidence for the Committee of which he was a member, stated its effect as follows:

1. French officers and those who witnessed the campaign from the French side were generally in favour of the employment of mitrailleurs in the field.

2. The Prussian Staff disapproved of their introduction into the service.

3. English officers who were present with the German Army, with one exception, considered that for certain purposes they might prove useful adjuncts to artillery.

In the experimental trials there was competitive firing at various ranges and under various conditions between Gatlings of ·45 and ·65-inch calibres, a Montigny machine gun, 9-pounder M.L. and 12-pounder B.L. field-pieces firing shrapnel and segment shell, and small squads of riflemen using the Snider and the Martini-Henry. The ranges were from 300 up to 2,100 yards. As might be expected, the machine guns heavily outscored the field-pieces at the shorter ranges up to 1,000 yards; and the Gatlings, in some cases, got more hits than the guns, even at the longer ranges. In one trial the large-calibre Gatling scored 164 hits at 2,070 yards, against 115 hits made by a B.L.R. 12-pounder firing shrapnel. Throughout the Gatling showed a marked superiority over the Montigny mitrailleuse.

In its report of 1871 the Special Committee adhered to its recommendation of the previous year, and proposed that the heavier Gatling of ·65-inch bore should be adopted for coast defence and for the Navy, and the smaller ·45-inch gun of the same system for the Army. There was some hesitation about ordering further guns, for after the Committee had issued its report news came of the invention of a Swedish gun said to be superior to the Gatling. At the time it was known as the Palmcrantz. It was the gun subsequently known as the Nordenfeldt.

In its final report the Committee recommended that machine guns should be used purely as defensive weapons; that, in general, they should be entrenched and kept masked from artillery fire, but that "the field artillery should not be reduced by a single man or horse for the sake of substituting machine guns."

In the Ashantee War of 1874 some of the Gatlings purchased in the winter of 1870-71 were sent out to West Africa, but they were taken no further than the frontier river, the Prah. Captain Owen says that they were found to be too cumbersome and top-heavy for the narrow road, and expresses the opinion that machine guns are not adapted for bush warfare. Clearly the weapon was not yet understood, and it was a case of the gun being handicapped by an utterly unsuitable wheeled-carriage mounting. If the Gatlings had been carried on pack-saddles, with tripod mountings, they would have proved their value against the Ashantee rushes at the battle of Amoaful.

Notwithstanding the partially favourable reports of the Special Committee in 1870-71, the War Office for some years paid little attention to the question of machine guns. As has been already noted, the machine gun was introduced into the Army long after its adoption by the British Navy, and as the result of practical experience of the use of the gun with naval detachments landed for service with the troops in our wars in Egypt and the Soudan. The machine gun had also been used by local troops in colonial frontier campaigns, and at home the Volunteers had experimented with it at manœuvres. We have seen that military opinion in England was for some years divided as to the value of the new weapon, and even after its value in our frontier wars had been recognized there was considerable opposition to its general adoption as a service weapon for our land forces.

From a lecture delivered at the Royal United Service Institution in June, 1898, by Lieutenant-Colonel Lockyer, we learn that it was in 1875 that the Gatling was definitely accepted as an army weapon, and in 1878 Nordenfeldt guns of three, five, and ten barrels were taken into the service.

There was still some hesitation as to the adoption of any machine gun on a large scale, and only moderate numbers of guns were purchased. In 1881 a committee of officers of both services made further investigations into the question, and carried out competitive trials at Shoeburyness, in which the guns tested were Gatlings, a two- and a five-barrelled Nordenfeldt, two American Gardner guns, and a gun of the "improved Gardner" or Pratt-Whitney pattern. One of the Gardners and a Pratt-Whitney gun were fired from tripods, all the others were mounted on wheeled carriages. A result of these trials was the adoption as service weapons in 1882 of the Gardner gun of three patterns, with one, two, and five barrels respectively.

The Army had now a certain number of machine guns of three different systems. But these were regarded somewhat in the light of exceptional appliances for special uses. They were mostly intended to be used for the defence of fortified posts. They were not an ordinary service weapon, with a recognized place in the fighting line. From some remarks made by Lord Wolseley in the discussion on a paper read by Major West, R.H.A., at the Royal United Service Institution on November 13, 1885, one gathers that for some years before this date the authorities at Woolwich had by their opposition to the proposal seriously delayed the adoption of machine guns by the Army. This opposition probably arose from the idea which prevailed for so long a time that the machine gun was intended to be used as a somewhat inefficient substitute for light field artillery. If this had been the case, the opposition would have been fully justified.

It was at this meeting of the Royal United Service Institution that Lord Wolseley, then Adjutant-General of the Forces, announced that the authorities had at last decided on making the machine gun an army weapon. At the same time he spoke of his conviction that there was a great future for the gun as a new arm for the infantry.

Despite this announcement, there was considerable delay in the general issue of machine guns to the Army. This was probably the result of official hesitation as to the best form of gun to be adopted. So far a certain number of guns of various patterns, originally manufactured for the Navy, had been handed over to the troops for use at certain stations in India and Africa. Thus the first Gatlings were in the hands of our soldiers as early as 1875. A number of Nordenfeldt guns were issued in 1878, and Gardner guns in 1882.* But in 1885 the Maxim had just been invented, and Lord Wolseley had been greatly impressed by the powers of this the first of automatic machine guns. The new principle seemed not unlikely to render all the older guns obsolete at an early date. Under these circumstances, some delay in the decision as to the type of guns to be adopted was natural and prudent.

One can see proof of the fact that the Army in those years depended for its machine guns on the Navy, in the circumstance that the only official handbooks for machine guns were those issued by the Admiralty. It was not until 1890 that the general issue of machine guns began, and the gun selected was the Maxim. Handbooks for its use were shortly issued by the War Office.

Up to this time the reports of the Hythe School of Musketry showed that instruction was being given there in

* These dates are given on the authority of Lieutenant-Colonel W. N. Lockyer, and are taken from a lecture which was delivered at the Royal United Service Institution on June 29, 1898, "Personal Reminiscences of the Evolution of Small Arms and Machine Guns from the Year 1863 to the Present Day." The position of the lecturer, who was then Chief Inspector of Small Arms, gives great authority to his statements.

the use of the older types of machine guns. In the report for 1888 we find it stated that the instruction in the use of rifle-calibre machine guns had been made an ordinary subject of the musketry course, and was tested in the examinations. Statistics are given of the number of officers and N.C.O.'s who had passed the qualifying examination as instructors in the drill and mechanism of various guns, the guns mentioned being Nordenfeldts and Gardners.

In the Hythe report for 1889 it is stated that experimental work had been done on the ranges with the Maxim gun, and for the first time instruction in the use of it had been given in the School, though the chief instruction work was still done with the Nordenfeldts and Gardners. These appear to have been regarded by the Hythe staff as the guns that would soon be generally issued to the Army, for the report adds:

> A manifest interest in the instruction of these guns is still exhibited by all ranks, and it is satisfactory to know that on their introduction into the cavalry and infantry there will be an ample supply of efficient instructors available.

In the report for 1890 we find the Maxim and Nordenfeldt named as the guns used in the ordinary courses at Hythe, "instruction in the Gardner being given to those belonging to corps in possession of that gun." The Gatling was still in use in India, and we find in the Hythe reports an account of field-firing tests made with it at Rawalpindi. In 1894, though the Maxim had become the standard army machine gun, instruction in the Nordenfeldt and Gardner was still being given at Hythe.

The issue of the gun to the Army, decided upon in 1887 and actually begun in 1890, went on very slowly. The first War Office Manuals for the Maxim were published in 1893, but it was a long time after this before every unit had received one Maxim gun, and this dearth of weapons meant defective training. Thus we read in the Hythe report for 1897:

Machine-gun practice is not satisfactorily conducted, and this can hardly be looked for until each regiment and battalion has been supplied with the gun. The present system works badly, and the officers, N.C.O.'s and men of the gun detachments very soon become rusty and forget all they have been taught on account of months—often years—passing without ever handling one of these intricate weapons.

The plain fact was that outside a group of keen enthusiasts who recognized the value and foresaw the coming importance of the weapon, there was not much interest in the machine gun in the Army. It was still regarded as very useful in fighting a crowd of savages, but a somewhat puzzling addition to the machinery of a battle line in civilized warfare. In the years immediately before the South African War every battalion and every cavalry regiment was provided with one Maxim gun mounted on a wheeled carriage. (A second Maxim gun was kept in every battalion mobilization store, and in 1910 this mobilization Maxim gun was allowed to be used for training purposes.) A few men were instructed in its mechanism and drill, but there was little opportunity for firing it, and this was generally done in a perfunctory way on a level rifle range against targets utterly unlike anything fired at in battle. The system of command was bad. A young officer was given charge of the machine-gun section for awhile. He had probably gone through a course at Hythe in Musketry, in which some instruction in the use of the Maxim was made an incidental part of a six weeks' course mainly devoted to the service rifle. He commanded his machine-gun section for a short time, then was promoted to a company, and unless he was an enthusiast dropped all interest in the machine gun and forgot most of what he had learned.

It is no wonder that in the South African War the machine guns gave disappointing results. A Hythe professor, in the résumé of his course of lectures issued to his class some years after the war, summed the matter up by saying that the comparative failure of the machine gun in South Africa was due to "want of knowledge of tactical handling and a low standard of training of the personnel. Other reasons were unsuitable targets offered by the Boers, and the carriage mounting." He was right on all these points except the question of the lack of suitable targets, for with good training, sound tactical ideas, and a suitable mounting, the targets would have been found. Some exceptional officers did very well with their guns, and some of these showed real resource in devising "dodges" for misleading the Boer gunners. Probably the professor was quite justified in adding that in official Army circles for some time after the war the machine gun "fell into further disrepute."

But things soon began to improve. The Manchurian campaign demonstrated the value of the machine gun with large organized bodies of troops. The revival of interest in it throughout Europe influenced our own officers. Special courses were begun at Hythe, and certain more advanced machine-gun work was carried out there. About the year 1909 the equipment of the machine-gun sections was improved. They were given two guns for training instead of only one. The Mark IV. tripod and the limbered waggon were issued to the Regulars, while the Territorial Force still had to train with the old Infantry Carriage Mark III., with its four-feet eight-inch wheels.

A considerable increase in the machine-gun armament of the army in France and Flanders was decided upon, and as it was impossible to obtain rapidly a sufficient number of machine guns, and it was recognized that other types of guns would also be useful, Hotchkiss and Lewis guns were purchased in large numbers. The Canadian contingent brought Colt guns with it from Canada, and the effective use they made of them demonstrated the value of this type of gun for certain work. The machine gun also proved its value as a weapon for the Armoured Motor Car squadrons of the Royal Naval Division.

In his account of the policy of the new Ministry of Munitions Mr. Lloyd George told how several factories in England had been equipped for the manufacture of additional machine guns for the Army, and as an instance of the use that might be made of them, stated that on one occasion the Germans held a position for days against a French division, with forty or fifty machine guns and a few hundred men as the only force that defended it.

But the most important step taken by the War Office was the formation on October 22, 1915, of a Machine Gun Corps, which is to be a corps for the purpose of the Army Act. This was the complete recognition of the machine gun as a new arm of the service, and the necessity for a special corps for the production of highly trained officers, N.C.O.'s, and men. At last a sound chain of command has been established, and officers can now devote the whole of their service to this subject, instead of only a few years while a lieutenant. By this means a large body of officers will be created, who can talk their own "*shop*" and learn the *lore of fire tactics* from the success and mistakes of their fellows. The corps should soon have its own journal and institution.

It should be noted that on November 12, 1914, a Motor Machine Gun Service was formed, one battery of which was added to each infantry division. In the first case the Motor Machine Gun Service was attached to the corps of the Royal Horse and Royal Field Artillery, and the ranges of the National Rifle Association became the training centre. The Machine Gun Corps is divided into cavalry of the line, infantry of the line, and Motor Machine Gun Service. These three branches will be under the Machine Gun Corps Record Office, but the pay will be respectively cavalry, infantry, and artillery. The branches will consist of squadrons, companies, and batteries.

This special recognition of the machine gun had long been urged by the small group of specialists to whose enthusiasm and perseverance most of the progress of machine gunnery in our Army was due. It is unfortunate that all these steps were not taken in the years before the Great War, but it must be granted that a laudable amount of energy has been shown in making up for lost time.

CHAPTER VII

MACHINE GUNS IN THE GERMAN AND AUSTRO-HUNGARIAN ARMIES

WE have seen that a few machine guns had been used with the Baden contingent in the war of 1870-71 without much practical result, and that the failure of the French machine guns in the same war for many years discredited the whole idea of the adoption of the new weapon in the German Army. Soon after he invented his gun Sir Hiram Maxim showed it in Berlin, and the Emperor William seemed to be greatly impressed by what he saw, and spoke enthusiastically of the possibilities opened out by the invention of the automatic gun. But even the German Kaiser cannot introduce any important change into the Army against the unanimous opinion of the General Staff, and for some years the Staff persevered in its theory that the machine gun was useless in European war. It was, however, adopted for the German colonial troops in Africa. It was admitted that it might be of use against ignorant savages whose tactics were simply an attempt to mob an enemy by a reckless charge in a mass. But the General Staff did not yet see how machine guns could be fitted into the more complex system of European tactics.

It was the Emperor's personal influence* that finally led to experimental trials of the new gun at manœuvres.

* In 1887 the Emperor, then the Crown Prince William of Prussia, had come to England on the occasion of Queen Victoria's first jubilee. Accompanied by a number of German cavalry officers, he paid a visit to the 10th Hussars, then at Hounslow, and was greatly interested in their machine-gun equipment, which then consisted of a

In 1899 Maxim batteries of four guns each were tried at the Imperial manœuvres as auxiliary weapons for the Jaeger battalion, attached to the cavalry. This was the first step to their general introduction into the German Army. At the same time the Emperor William, at his own personal cost, provided a machine gun of the same type to each of the Dragoon regiments of the Guard. In the following year machine-gun batteries were attached to the cavalry, and at the manœuvres of 1901 the experiment was made of employing machine guns with detachments of cyclists in connection with the cavalry division.

German military opinions were at first divided on the subject of the new departure. Writing in the *Jahrbuch* (December, 1901), General Rohne said that no one knew exactly what to do with machine guns, and the Commanders of army corps at the manœuvres who were given batteries of them, found themselves rather embarrassed by the new weapon. Probably, he said, it would be found that they were more useful in the defence than in the attack. Several writers, including some of the Generals, said the machine gun was far from being as effective in its fire either as the rifle or the field-gun. Captain Fritz Hoenig, who had a well-deserved reputation as a military critic, gave it as his opinion that it would be prudent to be very sparing in the supply of machine guns to the field army, and that one battery for each army corps would be quite enough. The military critic of the *Kölnische Zeitung* was more optimistic and far-seeing. He gave it as

Nordenfeldt machine gun on a galloping carriage drawn by two horses. The Crown Prince at once ordered a gun of the same kind to be sent over to Germany, and Colonel (afterwards General) Brabazon, then commanding the Hussars, sent one of his machine gunners to Potsdam to train the German troopers in the working of the gun. (*Journal of the R.U.S.I.*, 1888, p. 968.) This is another proof of the Emperor's keen personal interest in machine guns in the years when the General Staff at Berlin was still opposed to their introduction into the German Army.

his opinion that the experience of the manœuvres showed that the machine gun would certainly be of the greatest use to the advanced cavalry of an army, rendering it, to a certain extent, independent of infantry support.

In the German Army, however, changes in armament and method are not much influenced by Press discussion. The group of specialists who form the General Staff base their decisions on the reports of what is being done by other countries, and on experimental work carried out under their own direction. In these first years of the century they were carefully working out for themselves the question of the use to be made of the new weapon, and were gradually extending the machine-gun organization of the Army. In the summer of 1901 five machine-gun detachments were definitely established, one of which was attached to the Guard corps, and the rest to other army corps. In 1902 a machine-gun battery was attached to every cavalry brigade in the frontier districts, and eight other batteries were formed, to be attached to army corps in the interior of the country.

Experimental trials had been made with various types of guns, and a slightly modified form of the Maxim had been adopted as a standard weapon. Its mounting was a distinctly new departure. Both the wheeled carriage of the field-gun type and the tripod were rejected in favour of a kind of jointed sledge. The sledge could be adjusted so that it could be easily fired by a man standing, seated, kneeling, or lying down. It could be drawn over the ground by two men. The gun was mounted on a pivot on the sledge, so that it could be given a sweeping fire over a large arc of the circle. Sledges of the same type were used for bringing up ammunition. On the line of march, guns, sledges, and ammunition were carried either on light-wheeled waggons or on pack animals. Fifteen thousand rounds of ammunition were provided for each gun. The gun detachment was made up of four men, and with a well-drilled detachment the gun could be mounted and brought into action in a quarter of a minute. In case of emergency the guns could be fired from the waggon.

At first the batteries had been made up of four guns. In 1903 the number was increased to six. According to a regulation, issued in March of that year, the organization of a battery was as follows:

One captain and three lieutenants, each lieutenant commanding a section of two guns; thirteen non-commissioned officers, one of these was an armourer, and each of the others was in charge of a gun or an ammunition waggon; sixty-three men; eighteen saddle and thirty-six draught horses; six guns. Officers, non-commissioned officers, and drivers were armed with sword and pistol, the rest of the men with the carbine.

The Emperor showed himself strongly interested in the new development and an enthusiast for the machine gun. The military Press reported many sayings of his in praise of the new Maxim batteries. One cannot always rely upon these newspaper reports, and there was probably some exaggeration in the story that after the Metz manœuvres in 1902 he said that a single brigade with plenty of machine guns to help it could hold an army corps in check.

At first the machine guns had been attached only to the cavalry and to the Jaeger battalions, intended to act with them. In 1905 the question of extending their use to the infantry was taken into serious consideration, and the reports from Japan resulted in the decision being made to form infantry batteries. In the same year experiments were made with armoured motor cars, armed with machine guns. Sixteen machine-gun batteries, of six guns each, were engaged in the Imperial manœuvres of that year.

The introduction of machine guns into the infantry led to the development of two kinds of organizations, roughly corresponding to those of the horse and field artillery. In the detachments for service with the cavalry all the waggons and carriages had four horses, and the men not provided with seats on the carriages were given saddle-horses. In the infantry detachments the carriages had only two horses, and the men mostly marched on foot.

Detailed reports of the extensive use made of machine guns in the Russo-Japanese War, especially by the Japanese, and the important results thus obtained, were published not only in the German technical Press, but by the military correspondents of the leading newspapers. This led to an agitation for a considerable increase in the machine-gun armament of the Army. In 1907 and the following year large numbers of machine-gun detachments were organized. In July, 1907, the establishment of a machine-gun detachment or battery was fixed at four officers and 87 men. There were seven guns to a detachment, six to be brought into action, the seventh a reserve gun, intended to replace a machine gun so badly damaged that it could not be immediately repaired.

The foot batteries of machine guns were first organized in the army corps on the French frontier, but by the end of 1908 every German regiment of infantry had its battery of six guns. As a German regiment consists of three battalions and is almost equivalent to a British brigade, the proportion of machine guns amounted to a section of two for each battalion, though their permanent grouping in a battery placed them at the disposal not of the battalion commander, but of the colonel of the regiment. It was much as if in our own Army all the machine guns of a brigade were grouped in a battery under the direct orders of the brigadier.

The year 1908 is an important date in the development of machine-gun armament of the German Army. In the January of that year fourteen millions of marks were allotted for new trials of weapons, and it was stated that the money would be chiefly spent upon experimental work with machine guns.

As the result of these trials various improvements were introduced. The machine gun and its mountings were greatly lightened, and Carl Zeiss, of Jena, provided for experimental purposes a number of prismatic telescope sights. The new guns were painted a greenish-grey—"Feld-grau"—the grey colour of the German war service uniform.

Before these changes the weight of the machine gun had been 26 kilogrammes, about 52 lbs. It was reduced to 16½ kilos. The sled or carriage from which it was fired had weighed 56 kilos, the improved carriage weighed only 24. The weight of the new gun and its sled together were 32 lbs. less than the weight of the old carriage without its gun. These changes were made in view of the method adopted of bringing the guns into action—namely, taking gun and sled from the travelling carriage and then dragging them by hand along the ground, or where the detachment was not under close fire, lifting them and carrying them by hand.

During the years in which the machine-gun armament of the field armies had been thus gradually extended and improved large sums had been voted in the annual budgets to increase the machine-gun armament of the fortresses.

With reference to the machine-gun developments in Germany in 1908, there is reason to believe that German tactical ideas were influenced to a considerable extent by some valuable work done at our own School of Musketry at Hythe. The late Colonel N. R. McMahon, of the Royal Fusiliers, the chief instructor of the school, had been elaborating a system of machine-gun tactics which is now to be found in our Field Service Regulations and Training Manuals. While these Regulations were still in the draft stage they were communicated to the German military attaché in London in exchange for other official information, and thus reached the German General Staff, which practically embodied them in their own Field Service Regulations. It would not, however, be quite correct to say that the German methods of using machine guns in the present war were entirely the outcome of this adoption of British ideas. They built upon the foundations thus obtained. The German General Staff is singularly open to useful suggestions from any quarter, and any officer in the German Army who has a new idea has no difficulty in having it examined and experimentally tested, and if it has any value it is soon put into practice for the whole Army.

Perhaps the most important point in the new development, dating from the year 1908, was in the rapid multiplication of the number of machine guns available. Once its value was realized by the German General Staff, it was decided that it should be freely used and in enormous numbers. It was not to be regarded as the gunner regards a piece of artillery. Unless under exceptional circumstances, to lose a gun is something disgraceful. But the Germans count the loss of a machine gun as an incident which must frequently occur, and regard it as of little more importance than the continual loss of rifles in every battle. They disregarded the old idea that the machine gun was a weapon to be brought into action occasionally, when some opportunity arose for its use. It was to be like the rifle—one of the ordinary battle weapons.

The machine guns were therefore turned out in large numbers. An officer of the General Staff was appointed to act as the chief of a new inspectorate charged with the development of the new arm, and a special school was created for officers, sergeants, and armourers in order to provide skilled instructors, and to have an adequate number of men to work the thousands of guns which the factories were producing. As to the number of these available at the beginning of the war, it is said that besides those allotted to the fortresses, the German Army possessed in August, 1914, some 50,000 machine guns. It is not possible to verify the statement, but the lavish use made of the gun during the war, the numbers that have been brought into action in every engagement, make the statement highly probable.

A German fighting-line, either in the open or entrenched, simply bristles with machine guns. In entrenched positions there are comparatively few in the front line. The Germans recognize that this line is singularly vulnerable to artillery, and it may be said that it is little more than the outpost line of the entrenched position. The machine guns are much more numerous in the second line,* and especially in the fortified villages and the redoubts which form its main strength. In the attack a somewhat similar principle is adopted. There are a few machine guns with the first firing-line, but they are brought up in whole batteries with the second line, which is intended to drive the attack home after the first has borne the brunt of the defenders' fire.

The Field Service Regulations (Felddienst Ordnung) of the German Army that are now in force were issued in March, 1908, just at the time when the new development of machine gun in the Army was in its most active phase.

In these Regulations there are frequent references to the organization and employment of machine guns in war. It is noted that where the cavalry is pushed forward to secure important points on the line of march, block certain approaches to the enemy, or keep defiles open, effective support can be given to the cavalry by sending machine guns with them.

* The following is an extract from the letter of an officer describing the attack on the German trenches near Loos on September 25, 1915:

"There was a hurricane of yells as my men raced across, and we reached the first German trenches in double quick time.

"A hot fire had met us as we plunged across, but it could not even cause a temporary halt, although men dropped pretty quickly.

"When I reached the German trench I could see that it had been practically blotted out by our gun fire.

"It was the 13th Bavarian regiment we were charging, and the men were throwing down their arms and surrendering all along the line when we got there.

"We did not stay there, of course, but made off for the second line. Then the full blast of the German machine-gun fire met us, and I had only gone a few yards when I got one through the shoulder.

"I dropped, but as I fell I could see my gallant fellows had already reached the second line of trenches, and were pushing their way through there. . . ."—(Published in the London *Evening News*, September 28.)

The transport of a machine-gun battery is defined as consisting of, for the first line transport, five led horses and six spare horses, and for the second line transport a four-horsed store-waggon, a two-horsed baggage-waggon, a two-horsed supply-waggon and a two-horsed forage-waggon.

The ammunition-waggons are not counted as transport, but belong to the fighting portion of the machine-gun batteries.

As to ammunition supply in action, it is directed that fully loaded sleighs of ammunition (six boxes each) are to be drawn forward into the fighting-line, empty sleighs, boxes, and belts are to be sent to the rear and refilled as soon as possible from the ammunition-waggons; as these waggons are emptied, they are to be refilled from the nearest infantry ammunition column.

As to the tactics of machine guns in action, the following directions are given in paragraphs 581-584 of the Regulations:

The fire effect of machine guns is influenced principally by correct sighting, possibility of observation, size, and density of the target, and methods of fire. It is further affected by the suddenness with which the fire is opened, by the number of machine guns firing at the same target, and by the enemy's fire.

The high rate of fire concentration of the bullet-sheaf, and the possibility of bringing several machine guns into action on a narrow front, enable great effect to be produced in a short time, even at long ranges. When the front of the target is broken and irregular, the effect is reduced. A wrong sighting elevation, or imperfect observation of fire, may render the fire completely ineffective.

Dense lines of skirmishers standing suffer heavy losses at ranges of 1,650 yards (1,550 metres) and under. At lines of skirmishers lying, good effect is to be expected at 1,100 yards (1,000 metres) and under, provided that the observation of fire is good.

Against artillery in action the fire effect is similar to that of infantry. Owing to the mobility of the machine-gun batteries they are especially adapted for securing the increased fire effect due to oblique fire.

At short ranges under hostile fire, machine guns can only be brought up and withdrawn under cover.

Machine guns, even when on their wheeled carriages, are able to emit a large volume of well-aimed fire in a short space of time.

In the notes on infantry fire it is pointed out that against machine guns limbered up on their travelling carriages, the fire effect is the same as against artillery in the same position. But against machine guns on the move on their sleighs, the effect is the same as against skirmishers, and it is added that machine guns in action form a difficult target, and their fire power may be very little affected even after part of the detachments serving them have been disabled.

This is the summary of the instructions as to machine guns to be found in the Field Service Regulations. But as the Regulations deal with the action of the whole Army and every branch of it in war, only the essential points are noted in dealing with each arm. There are besides Special Regulations for the machine gun arm, supplemented by orders and instructions issued from time to time. A translation of these Regulations for the information of the French Army by Captain Shoenlaub has been issued in Paris by Charles-Lavauzelle since the declaration of war. The latest instructions it contains date from 1912.

These Regulations show that by that date the machine-gun organization of the German Army included two different kinds of machine-gun batteries—the machine-gun company and the machine-gun detachment (*Abtheilung*). Machine-gun companies had been attached to every regiment of infantry in the Army. The German infantry regiment is organized in three battalions of four companies each, numbered throughout the regiment from one to twelve, and the regiment has nearly the strength of a British brigade. The machine-gun company in each regiment is numbered 13. There are six guns in three sections of two each, and the company is directly at the disposal of the Colonel. He can either use it as a six-gun battery, or work the section separately—for instance, sending a section of two guns with each battalion.

The machine-gun detachment, also of six guns in three sections, is non-regimental. It is an independent battery of machine guns. The number of these detachments has been enormously increased since the war began. Detachments of machine guns, with the officers and non-commissioned officers and a number of the men mounted, and the rest conveyed on the carriages, are attached to the cavalry divisions. Motor-car detachments have also been organized by using these cars instead of horse transport, and mounting the guns so that they can be fired from them.

The Regulations note that besides the six guns attached to the machine-gun detachment, it has in time of peace two older guns, which are used for instruction, so as to minimize the wear and tear of the service pieces. Elaborate instructions are given as to how all the material of the battery is to be kept in good order.

The Regulations for the preliminary drills include exercises in taking cover and handling the machine gun in various positions—standing behind a parapet, kneeling, seated, and lying down. The first firing exercises are with blank cartridges. Two kinds of firing are distinguished—continuous firing, and salvo, or volley-firing, this latter being a discharge of fifty cartridges, usually employed for range-finding purposes. Continuous firing is again divided into three varieties: (1) Fire concentrated on a single point, (2) firing with a horizontal sweeping movement, and (3) firing with continual change of elevation, so as to give depth to the beaten zone on which the bullets fall.

After the preliminary drills the detachment is exercised in target practice on the range and field-firing. It is directed that in all practices care is to be taken that the same men fire the same gun, so as to become familiar with any peculiarity in its shooting. There is a direction which one expects in case of artillery practice, but which seems hardly needed with machine guns. At all firing practices, one of the non-commissioned officers is responsible for having a supply of cotton ready to be put in the ears of those working the guns.

The range practices include firing at targets up to 1,500 yards, with slow or rapid fire, concentrated and sweeping fire, fire with varying elevation, and with the machine gunner kneeling, seated, or lying down. Elaborate instructions are given; and there are special tests for second and first class shots, and those who pass these tests qualify by an additional test as pointers.

The field-firing practices are divided into firing with a single machine gun, firing with the section of two guns, and firing with the whole battery of six guns. It is directed that these practices shall take place on at least twelve different days and at various seasons of the year. Everything is to be carried out as far as possible under battle conditions.

Great stress is laid on the choice of targets, and of the moment for coming into action, and it is noted that the heavy expenditure of ammunition and the wear and tear of the material restrict the time during which the guns can be kept firing, and therefore make it all-important that the fire should be efficacious. Oblique or flank fire is the most effective against troops of all kinds, and the only kind of fire that will give much result against artillery with shields on the guns. As it may be difficult to pass orders by word of mouth amid the noise of battle, officers and men are to be practised in giving and obeying orders by signal.

The inspection of a machine-gun company or detachment is not to be confined to mere parade movements, but is to include actual firing. Arrangements are also made from time to time for experimental firing under the direction of the General Staff, in order to obtain data for the solution of questions that it is studying.

Prizes are given by the Government for special competitions in shooting with the machine gun at the end of the annual course. For instance, the best shots among the non-commissioned officers at any military station are allowed to compete in target practice, in which each fires 250 shots; the prize is a watch given in the name of the Kaiser, and engraved with the name of the winner.

The usual machine-gun targets are bands of canvas, or cardboard, representing the heads and shoulders of lines of skirmishers. These are placed either at right angles or obliquely to the line of fire, or several targets are arranged behind each other to represent successive lines of skirmishers.

There is a liberal allowance of cartridges; each company or detachment is annually allotted 110,000 ball cartridges and 100,000 blank cartridges, fitted with a dummy bullet of soft wood, or compressed paper, so as to produce sufficient recoil to work the breech mechanism. It is laid down that 70,000 ball cartridges are to be expended in the field firing, and 7,500 at the annual inspection.

It is forbidden to economize in any way the 70,000 cartridges allotted to the field firing. If by any chance it is not possible to use them all in a given year, the unexpended cartridges are to be carried forward to the following year and added to those used at the field firing.

The 100,000 blank cartridges are to be used thus: 40,000 for drill and training purposes, and the remaining 60,000 at the annual manœuvres.

Each soldier of the machine-gun company or detachment has a book, in which a record is kept of his shooting at every practice in which he takes part. The commander of the company has thus a permanent record showing the relative efficiency of all his men.

When a company or detachment receives a new gun, its shooting is carefully tested on the range, 1,000 cartridges being allowed for this purpose. A diagram of the shooting is made, and this is kept as a guide in the subsequent use of the gun.

So far the Regulations deal with the training of officers, non-commissioned officers and men. There are further manœuvre Regulations which include instructions for the actual fighting of the guns. An introductory note sets forth that:

> The essential point for the machine-gun detachment is to be able to shoot well and open fire at the opportune moment from the most favourable position, and against the most useful objective. To secure all this, it is necessary to have a thorough knowledge of the gun, a very mobile detachment, leaders who possess tactical insight, and men full of initiative who, devoted heart and soul to the Emperor and the Fatherland, will exert themselves to obtain victory, even when they have lost all their leaders.
>
> The machine-gun detachments will be of no use in battle unless the men are accustomed by frequent practice to take full advantage of the ground, to choose their position judiciously, to estimate and measure distances accurately, and to get the range quickly. And they must know the tactical methods of the other arms, especially of the infantry. Manœuvres with other arms of the service are therefore to be insisted upon, for these alone enable one to realize the difficulties one has to encounter in the handling of machine guns, and learn how to surmount them.

It is pointed out that in selecting a position and coming into action, the first thing to be thought of is the best fire effect, rather than the mere avoidance of loss.

Machine guns can be employed on any ground practicable for infantry, and when they are taken from a travelling carriage they ought to be able to pass even difficult obstacles. The target they present to the enemy is not greater than that which is offered by men in skirmishing order. They are more difficult to put out of action, and they give their commander the power of bringing an intense fire to bear upon any selected point.

Any cover that is sufficient for infantry will be enough for a machine gun, and cover that would hardly give room for an infantry section (a *Zug*, or one-third of a company) will be sufficient to shelter a whole battery of machine guns.

Machine-gun detachments can also keep up with and come into action with mounted troops.

The rapidity of their fire, the concentrated sheaf of bullets, and the possibility of bringing several machine guns to bear upon a small space of ground, give one the possibility of obtaining decisive results at a given point, and even of annihilating at long range targets which have a wide front or considerable depth.

Machine guns come into action at the favourable moment for a comparatively short time. They are not so well fitted for a prolonged fire fight.

As a rule one should avoid engaging well-sheltered firing-lines of hostile infantry. A fight of this kind requires an expenditure of ammunition out of proportion to the results obtained. It will therefore happen that in battle the machine guns will remain for a long time inactive in their firing position, reserving themselves for the decisive moment.

To put the enemy's machine guns out of action, artillery fire is more effective than that of one's own machine guns. At extreme ranges the artillery has a marked superiority. If opposed to artillery, the machine guns ought to be pushed as far forward as possible, and thanks to the mobility, they can try to reach a flank position and bring an oblique fire to bear upon the guns. Frontal fire against artillery in action does not produce much effect.

A battery of machine guns has nothing to fear from the enemy's cavalry. If they attack, any formation is good which allows the machine guns to bring a rapid fire to bear upon the charging cavalry. But the fire must be delivered calmly and with precision. Even on open ground, machine-gun batteries should be able to repulse cavalry, provided these are not so numerous that they are able to attack in several lines and in several directions at the same time. It is sometimes useful to attach mounted scouts to the groups of machine guns, but it is seldom necessary to give them an escort. This may be required in very close country. The escort will be made up of a small detachment of infantry or cavalry, and its business will be to watch the flanks and rear and protect the transport of the battery.

As a rule the company or detachment of machine guns remains united, but if the situation requires it, sections may be detached. It is, however, forbidden to detach an isolated machine gun. Two guns must always come into action together. There will rarely be any advantage in concentrating a number of batteries of machine guns. If, however, this is done, the senior machine-gun officer takes command of the whole.

At the beginning of an action a machine-gun commander goes to the officer under whom he is acting to receive his orders for the employment of the guns. If necessary, he will ask for orders. During the fight he will keep in touch with the officer in command. He is to give his own orders briskly and without hesitation, and he must remember that negligent inaction is a much graver fault than any mistake he may make in choosing his course of action.

In choosing a position he must think first of securing a good field of fire, and after that about cover. Before occupying the position there must be a reconnaissance, and the skilful execution of this is one of the essential conditions of a satisfactory result. The commander of the group of guns should make the reconnaissance in person, except during a retreat, when he will remain with his guns so long as they are under fire, and send the next senior officer back to choose the position to be occupied on retiring.

The reconnaissance to the front should be made on foot, in order to avoid attracting the attention of the enemy to the selected position. In choosing a position it is well to select one which enables the ground to be kept under fire to within a close range of the guns. It should afford facilities for movement to a flank or to the rear. It should not be too near marked objects that would help the enemy to get the range.

The background is important. Care should be taken that the guns do not show up against it.

While moving into position, all necessary precautions for security must be observed. Scouts should be sent out to an exposed flank, especially in close country. The road should be used as long as possible. The orders for taking up the position must be given in good time, so as to avoid any delay in opening fire. One should try to reach the position without being seen, and open fire by surprise. But even if there is not sufficient cover in the approach to the position, one will try to surprise the enemy by rapidity of action.

As a rule the interval between the guns will be about twenty paces, but one must not try to place them evenly in line at equal intervals. Each machine gun is put where it will have the best field of fire and the best cover. The vulnerability of the line of guns increases with its density. If a flank is in danger, it may be well to place the guns *en echelon*. When the nature of the ground or of the target require a careful choice of the position of each piece, it is well to bring on in advance the non-commissioned officer who commands each gun and even the man who will point it.

Care must be taken not to open fire prematurely. One must remember that fire will not produce a decisive action, unless it is directed at hostile troops within good range and affording a good target. Before opening fire, remember that the number of cartridges is limited, and the employment of a quantity of ammunition represents a loss of fire power which has to be made good as soon as possible. But once fire is opened, it must be kept up until the object in view is obtained. An inefficacious fire decreases the morale of one's own men and strengthens that of the enemy. The loss inflicted on the enemy will produce a greater effect in proportion to the shortness of the time in which it is inflicted. As a rule fire will be opened therefore with all the guns available. This is better than bringing one or two sections into action, the expenditure of ammunition will be much the same, the effect on the enemy greater, and one's own losses less.

The target will not be changed until the effect in view has been obtained. Frequent changes of target diminish the effect of the fire. Machine-gun fire at night is only effective if the guns have been pointed during the daytime on places where the enemy must pass, or if one has a well-lighted target, such for instance as bivouac fires.

With a well-disciplined detachment the guns should be kept in action, even if the officers are *hors de combat.* Persistence in keeping up the fire of the detachment can secure success, even after heavy loss, and in such a case it should be remembered that the enemy is probably suffering just as heavily and is under the same stress.

The waggons are kept under cover; directions are given for bringing up the ammunition-sleighs to the guns, and it is suggested that where the ground gives good cover, a heavily loaded sleigh can be brought up by one of the horses, a man walking beside it to steady it on rough ground. Empty cartridge-belts will be sent back, and refilled by the men with the waggons.

In the attack, machine guns are pushed forward with the firing-line. But they need not always be actually in it. If they can occupy a position at a range of 800 yards or more, from which they can support the advance by their fire, it is a mistake to push further forward. They can be usefully employed to protect the advancing infantry against flank attack, and they supply a reserve which the commander of the attack can use to bring a heavy fire suddenly to bear upon a decisive point. In the attack on an entrenched position, it is well to place the machine guns in position during the night, so as to be ready to open fire at daybreak. The guns will help to force the enemy to keep well down in his trenches, thus facilitating the assault.

On the success of the attack, machine guns will be rushed

forward to assist in holding the ground that has been won against any counter-attack. If the assault fails, the machine guns will help to cover the retirement of the attacking troops.

Machine guns can be of great use in the pursuit, on account of their coming quickly into action, and delivering a rapid fire. All available ammunition should be pushed forward with them.

On the defensive, machine guns can be partly held in reserve, partly assigned to special points on the line of defence, where they will be entrenched. They are especially useful to sweep with their fire lines of approach that the enemy is likely to follow, and to bring fire to bear upon dead ground in front of the works, and to guard the flanks.

In a retreat they can be used with advantage to hold positions in the rear of defiles, or to sweep open ground which the enemy is likely to cross.

Machine guns attached to cavalry acting independently will take their general directions from the cavalry commander. It is often useful to attach even a single section of machine guns to a cavalry detachment. When cavalry is in action against cavalry, the machine guns should endeavour to bring fire to bear upon the enemy from an advanced position on the flank. It can thus keep up its fire to the moment when the charge is driven home, and at the same time protect the flank of its own cavalry from a turning movement. It is not well to bring several sections into action separately in such a case, for their various lines of fire may hamper the movements of their own cavalry. The machine-gun commander should not wait for the orders of the cavalry commander. A cavalry action develops very rapidly, and he must intervene on his own responsibility.

In all cases machine guns will generally use direct fire, but occasions may arise when it is possible to use indirect fire, by means of marks placed in position in advance, or with the help of an observer.

Officers of all arms should have a general knowledge of the principles of the use of machine guns, so as to be able to act in concert with them.

With the cavalry divisions, machine-gun detachments can often be usefully employed with the horse artillery. In covering a retreat commanders of machine guns must not hesitate, if necessary, to hold on, even at the risk of their guns being captured.

* * * * *

German tactical methods during the present war have influenced the use of machine guns in the armies of Germany's ally, Austro-Hungary. But Austria had been in advance of Germany in the adoption of the machine gun.

Some trials of the Maxim machine gun were made in Austria as early as the year 1887, but it was not till 1893 that the machine gun was introduced as an auxiliary weapon for the Austrian cavalry and for the fortresses. The system adopted was the invention of the Archduke Charles Salvator, in collaboration with Major von Dormus. The gun was perfected and manufactured at the Skoda Works, and patented in the name of the Skoda firm. It was an adaptation of the Maxim principle, but had a slower rate of fire, about 300 shots a minute. In Austrian military circles opinion in favour of the new gun was not unanimous. Soon after its introduction the *Vienna Reichswehr* said that its only real advantage was that it was an Austrian product and could be cheaply manufactured.

The Skoda machine gun was not issued to the Army generally. Up to 1898 it had been supplied only to the fortresses and the cavalry regiments in garrison on the frontier.

A number of Maxim guns which had been purchased for experimental purposes were given to some of the cavalry regiments instead of the Skoda gun. But in the Austro-Hungarian Army both were superseded in 1906 by the Schwarzlose machine gun, a modified Maxim, with a greatly simplified action. It had been under experimental trials for about four years before this date. The chief improvement introduced into its action was the elimination of nearly all the springs from the mechanism. A spring is always a weak point in a mechanical device, the tension continually varies in use, and has to be carefully adjusted, and unless there is very efficient supervision the collapse of a spring may at any moment cause a breakdown. In the original Maxim there were fourteen springs. There is only one in the Schwarzlose. The barrel is fixed, and the recoil of the cartridge as it is forced out of the breech of the gun works the mechanism. To put the matter in familiar language, when a shot is fired the bullet is driven forward and the cartridge-case backwards, the latter, however, moving much more slowly because it at once encounters the resistance of the mechanism which it actuates. It is, however, essentially a device derived from the original Maxim plan, because it uses the recoil to work the mechanism, the difference being that in the one case the end of the barrel applies the force of the recoil, and in the other the base of the cartridge-case. The guns were made under contract with the Government by a company at Steyr in Austria. The Schwarzlose appears to be a very effective weapon. In the final trials two guns fired each 35,000 cartridges, without any serious repairs being necessary, and though the same barrels were used throughout, the fire continued accurate to the end of the trials. In 1906 special machine-gun schools of instruction for the Austro-Hungarian Army was established at Bruck on the Leitha. The organization adopted for the Austrian cavalry detachment was three officers, forty-five non-commissioned officers and men, and sixty horses, two guns carried on pack-saddles and four pack-horses carrying cartridges. On a war footing this last number would be increased to twelve. Mountain batteries of machine guns were organized, with six guns, in three sections of two each, the large bore thirty-seven millimetre Maxim gun throwing small shells was adopted for the fortresses.

According to a Vienna paper of 1907 the machine-gun detachments were provided for peace manœuvre purposes with a blank cartridge in which the bullet was replaced by a paper plug. This gave sufficient resistance in the barrel to produce a recoil strong enough to work the mechanism of the gun. The principle is the same as that of the blank cartridge with a soft wooden plug, which was used in the same way by some of the Volunteer regiments in England for their machine guns, and is in use in the German Army. All these arrangements have the drawback that there is some danger of accidents at short ranges through the plug not always breaking up.

Officers and non-commissioned officers attached to machine-gun detachments have to qualify by following a six months' course of training at Bruck.

The Austrian Regulations for machine-gun detachments issued from Bruck in 1907 estimate that the fire of a single machine gun is equal to that of a Zug or section of infantry, about 80 rifles. But it is pointed out that this comparison only holds true where the *guns are well handled and used under conditions that give the fullest play to their special qualities*. Special stress is laid upon the necessity of a high degree of training for every member of the detachment, and of a spirit of initiative in officers, section commanders and the non-commissioned officers in charge of each of the guns. It is pointed out that a number of the men in the ranks must be so well trained as to be able to take the place of the non-commissioned officer. There is a liberal allowance of ammunition for training purposes: 15,000 ball cartridges are issued annually for each gun, besides blank cartridges for manœuvre exercises. At the autumn manœuvres of 1907, 5,000 blank cartridges were allotted to each machine gun.

The first mountain detachment of machine guns was formed in Bosnia and in the Alpine garrisons on the Italian frontier. The Austrian machine-gun detachments are said to be very efficient. The School of Instruction at Bruck has for years been doing valuable experimental work, and experiments have also been carried out in the mountain districts in the use of machine-gun fire against various slopes of ground and against the crests of precipices. Many of the results obtained have been published in the technical reports of the Bruck School, of which about six are issued annually.

CHAPTER VIII

MACHINE GUNS IN VARIOUS FOREIGN ARMIES

IN the chapter on the evolution of the machine gun, something has been said of the introduction of the weapon into various foreign armies, and in a special chapter we have dealt with the machine-gun equipment of the German Army and that of Austro-Hungary.

It may be useful here to add some brief notes on the machine gun as used by other foreign armies.

As already noted, RUSSIA was the first European Power which adopted the machine gun as part of the normal equipment of its troops. The first machine gun used in the Russian Army was the Gatling, but before the Japanese War it had been superseded by the Maxim. There are two patterns, those of 1905 and 1910. They differ only in weight, the new gun having been considerably lightened, chiefly by substituting steel for gun-metal in various parts of the gun and its mounting. The 1905 gun weighs 63½ lbs., the 1910 gun only 40. Both these weights are with the water-jacket empty; when full it holds 8½ lbs. of water. There are both wheeled and tripod mountings; the standard mounting, designed by Colonel Sokolov, which during the last few years has gradually superseded all others, is a combination device with wheels, a trail and two folding legs, which can be placed so as with the trail to form a tripod for the gun, and can be adjusted in two positions, the first for firing with a gunner seated on the trail and the second with the gunner lying down. A steel shield weighing about 20 lbs. is fitted to the tripod.

A machine-gun detachment has eight guns organized in four sections of two each. A detachment is assigned to every Russian infantry regiment and every cavalry division. As a Russian regiment is made up of four battalions, this gives a section of two guns to each battalion, if the detachment is divided.

Both with the infantry and cavalry detachments the machine gun on its fighting carriage is conveyed on a two-wheeled cart on the line of march when action is not immediately expected. But it is so mounted on the car that on an emergency it can be fired from it. When taken off the cart, with the infantry it is drawn by one of the two horses that form the team, or it is drawn by hand; with the cavalry it is mounted on a pack-saddle.

The infantry detachment of eight guns has three officers and eighty rank and file. Two thousand rounds of ammunition are carried on each of the machine-gun carts, and there are besides 18,000 rounds carried on eight ammunition carts. The ammunition carried on the machine-gun carts for immediate use is packed in a peculiar way. There are eight belts of 250 rounds each, and these belts are packed in two metal cylinders, each of which thus holds 1,000 cartridges. When they are taken off the cart, the cylinders are rolled along the ground.

With the cavalry, in order to make it easier to travel at a quick rate, each machine-gun cart has four horses. There are, besides, three pack-horses—one to carry the gun when it is taken from the cart, the two others for ammunition. There is a non-commissioned officer and 13 men for each gun, all of them mounted. The total strength of the detachment is 5 officers and 113 rank and file, besides 18 non-combatants, in charge of carts. There are 198 horses. A Russian cavalry machine-gun detachment has, in fact, as many men and horses as a field battery in some armies.

Besides instruction with the troops to which they are

attached, all machine gunners go through a special course of instruction lasting a fortnight in each year. About 50,000 rounds of ammunition are allowed to each detachment for this course, which is carried out at a specially organized training centre for machine guns.

In the official manuals great stress is laid on the initiative to be left to machine-gun officers once their special task has been explained to them. The normal fire unit is two guns, but unlike most regulations, those of the Russian Army expressly state that guns may also be employed singly. The sections are usually to act separately, and the use of several sections as a battery in action will only occur under exceptional circumstances. In the infantry a section is usually allotted to each battalion in the firing-line, and it is unusual to withdraw guns from the regiment to which they are attached.

It is laid down that the commander of a cavalry detachment of machine guns should usually be with the cavalry commander to receive his instructions. The guns will try to act on a flank to support the action of the cavalry by indirect fire. In dismounted action they will be worked like infantry machine guns. The guns have no escort, but a few mounted scouts are generally attached to them.

With the cavalry detachments there are five officers for each, and this gives, besides the detachment commander, an officer to command each section. But the cavalry detachments are few, there being only one for each division. In the case of the much more numerous machine-gun detachments attached to the infantry, there appears to be a dearth of trained officers, for only three are assigned to each detachment, and the Regulations lay down that one of these will command the whole detachment, leaving two to command sections. The other two sections are to be commanded by sergeants. The arrangement is obviously defective, and can only be explained by supposing that sufficient machine-gun officers are not available.

In 1896 a number of Rexer guns were purchased for the

cavalry, and in the latter stage of the Russo-Japanese War numbers of these guns were supplied to the cavalry regiments in Manchuria, but the war came to an end before much use had been made of them. The Rexer is still in use in Russia as a cavalry weapon.

SWITZERLAND was one of the first countries to adopt machine guns. In 1889 the Federal Diet voted a small credit for the purchase of a few of these guns for experimental and training purposes. After a long series of trials it was decided to adopt the Maxim. Cavalry detachments were first formed, and batteries were then organized to work with the infantry in defence of the mountain passes.

After a period of experimental trials on the range and at manœuvres, the organization adopted was that of mounted companies attached to the cavalry brigades for service in the field, and foot companies as auxiliaries for the defence of fortified positions. The companies were practically batteries of eight machine guns organized in four sections of two guns each. In 1902 four mounted and three unmounted companies had been formed. The mounted companies were attached to the cavalry brigades of the field army. Two of the unmounted companies were attached to the defences of the St. Gothard—the system of fortifications commanding the St. Gothard Pass, the Schollenen Pass, the Furka Pass, and the mountain road that leads from Andermatt to the head-waters of the Rhine—fortifications that have converted the Andermatt Valley into perhaps the strongest fortress in Europe. The remaining unmounted company is attached to the defences of the lower Rhone Valley at St. Maurice.

The mission of these unmounted companies is to supplement the defence of the main passes by supplying flanking fire from the heights, and, further, to be ready to assist in closing against an enemy the many minor hill paths by which the attack might try to turn the permanent defences. The guns are conveyed on pack-animals, but can also be carried on ground where even mules could not make their way, as, for instance, over glaciers and icy slopes, by the gunners themselves. For this purpose the gun is mounted on a framework which can be strapped on a man's back, and from which it can be fired when placed on the ground. It is easy to find men capable of this heavy work of carrying guns and ammunition-boxes on their backs over difficult mountain ground. In Switzerland during the summer season, when the herds are out on the upper pastures, just below the snow-line, one sees young men, almost boys, coming down from the hills carrying heavy milk-churns on their backs. This is good preliminary training for carrying a machine gun in the same way. The unmounted machine-gun companies are practically Alpine troops, and can carry their guns wherever a Swiss mountain guide can make his way. During their summer training, they are practised in bringing the guns into action, even in positions above the snow-line.

The mounted detachments have both wheeled and pack-transport for their guns. At first sight Switzerland would not seem to be a country adapted for cavalry action, but north of the Oberland range the country falls rapidly to the northern plain, and a considerable part of Swiss territory is fairly open, undulating ground. But the Swiss cavalry are employed at manœuvres even in the higher ranges. The excellent roads that traverse the valleys and most of the higher passes make rapid mounted movements comparatively easy. Further, at manœuvres one sees the flankers of the advanced cavalry making their way over the mountain paths to right and left of the road. In this hill work the machine guns are attached to the cavalry in order to supply them with concentrated fire power to enable them to seize and hold commanding positions in a defile or pass until the infantry can come up. In work of this kind the cavalry are really highly trained mounted infantry, and the machine gun is admirably fitted to increase their fighting power. The Swiss Army Regulations lay it down that "machine guns" are attached to the cavalry for the purpose of increasing its effective fire power.

In FRANCE, for many years after the war with Germany in 1870-71, machine guns were discredited, as a result of the failure of the mitrailleuses. The Navy had adopted for action against torpedo boats and for use with landing parties a revolving gun with several barrels, throwing a heavy solid bullet or a small shell—the "*Canon Revolver Hotchkiss.*" It was the invention and was manufactured by an American engineer, Mr. B. B. Hotchkiss, who had a factory at St. Denis, near Paris, founded in 1875. Besides his revolver cannon, he had invented several systems of breech actions for artillery. His machine gun, useful as it was to the Navy, was much too heavy for land work, and besides this, if used with troops, it would have been at best a rather inferior kind of quick-firing artillery.

He was anxious, however, to devise a machine gun for land service of the rifle-calibre type, and he took up and greatly improved a gun designed by the Austrian Captain von Odkolek. In its improved form it was patented as the Hotchkiss machine gun.

It was a gun of the gas-engine type. Before the bullet leaves the muzzle a portion of the gas produced by the explosion escapes through a hole under the barrel a few inches short of the muzzle. This gas works a piston which operates the breech action and the ejecting and reloading mechanism. The ammunition is fed from left to right by means of a metal belt. There is no water-jacket. The gun is air-cooled, with the help of flanges on the barrel, and there is an arrangement for rapidly changing the barrel if it becomes overheated. It was claimed that the gun could fire at the rate of about 650 shots a minute.

In 1899 the Minister of War, General de Gallifet, introduced the Hotchkiss machine gun into the French Army. Two forms of Hotchkiss were adopted, both firing the Lebel cartridge, used by the French infantry, the only difference between the two types being that one was much lighter than the other, used a shorter cartridge-belt, and was practically an automatic rifle. The guns were supplied from the Hotchkiss works at St. Denis.

They had no sooner been introduced into the French Army than various suggestions were made for the improvement of the gun in matters of detail. Under an arrangement with the Hotchkiss Company, it was decided that one or more of the Government arsenals should undertake the manufacture of the weapon, and a number of guns were turned out by the State manufactory of small-arms at Puteaux. The staff of the arsenal introduced various changes into the design, and prolonged departmental and ministerial debates as to the adoption of these led to hesitation as to manufacturing the gun in any large quantities, and so retarded for a while its distribution in the French Army.

During these debates a third type of Hotchkiss was designed by the staff of the Government Arms Factory of St. Etienne. This suggestion appears to have led to further delays, the Ministry of War giving only very limited orders for the guns, because it could not make up its mind as to the standard type to be adopted. Meanwhile the operations in Morocco in 1907 had given abundant opportunities for testing the guns in the field. Hotchkisses of both the St. Denis and the Puteaux pattern were used by the French columns in their battles with the Moors, and the reports both of newspaper correspondents and of the officers in command were unanimous as to the services rendered by the new guns. It must, of course, be remembered that the conditions were different from those of European warfare. The Moors gave easy targets to the guns by their practice of attacking in masses and even attempting to charge on horseback.

The reports from the Moorish campaign, the articles published in the French military press on the important part that machine guns had played in the Russo-Japanese War, and the news that the German Army was being provided with an ever-increasing number of Maxims, led to an agitation in France for a prompt decision as to the type of gun to be adopted, and the rapid provision of a sufficient number of guns for the fortresses and the field army. M. Charles Humbert, a deputy who had for some years given much attention to military questions and taken a leading part in the debates on the army in the French Parliament, published in October, 1907, under the title of "Sommes-Nous Défendus?" a remarkable book which went through twelve editions in as many weeks. It was an outspoken attack on the administration of the French Army, and the experience of the present war has proved that many of M. Humbert's charges against the War Ministry, notably those referring to the neglect of the fortresses, were well-founded. A considerable part of the book was devoted to the question of the machine guns. M. Humbert declared that large numbers of regiments were without the new weapon, and that the provision of machine guns for the fortresses was sadly deficient. The four great fortresses of the eastern barrier ought, he said, to have at least a thousand of them, but they had only three or four dozen. He argued that if the rapid manufacture of the guns in large numbers was to be deferred until the technical advisers of the Minister of War had evolved an ideally perfect weapon, the result would probably be that when war came the French Army would find itself hopelessly short of machine guns. In the debates which followed in the Chamber of Deputies, the Ministry made the usual official reply to M. Humbert's attack, and obtained a vote of confidence from the House, chiefly because it promised to push forward the manufacture of machine guns both by the Hotchkiss Company and the Government factories, and there was the usual indisposition to produce a Cabinet crisis on a side-issue.

The agitation had the usual result that there was an end

of the official search for the ideal gun, and the manufacture of both the original Hotchkiss type and the St. Etienne model was vigorously pushed forward.

The French Army has thus two types of gun, but the differences between them are only in matters of minor detail. The official regulations now in force for the guns in the French Army were issued on November 25, 1912. The book "Réglement sur les Sections de Mitrailleuses d'Infanterie" is in two small volumes, the second of which relates entirely to the guns and their mountings, ammunition, range-finders, and material generally. The first volume is the book of Instructions.

The guns are organized in sections of two guns each; normally each section has partly wheeled and partly pack-transport, but in the 14th and 15th Corps (Corsica and North Africa), and in sections attached to the Alpine troops, there is pack-transport only.

In time of peace machine-gun sections of the First Line Army have an annual allowance of 15,000 ball and 20,000 blank cartridges, those of reserve regiments are allowed 5,000 of each. But even for the First Line Army the amount of ammunition is far below the German standard. When mobilized for war, the machine-gun sections are under the direct control of the army corps commander. In time of peace a section is usually attached to each battalion.

The normal rate of fire (*cadence moyenne*) is 200 to 300 shots per minute. The Instructions distinguish between *tires blocqués*, or concentrated fire on a narrow front, and *tires déblocqués* (sometimes described as *tires déblocquées avec fauchage*), when the gun is traversed.

In the Tactical Instructions it is pointed out that care must be taken not to attract the attention of the enemy while coming into position. The position should have both a good field of fire and cover, and to diminish loss the two guns should be kept well apart, but near enough for one officer to control both. The position should be a dominant one, or to a flank, so that the infantry advance will not mask its fire. Ranges are to be taken with the range-finder. Action at extreme ranges is generally not advantageous. The expenditure of cartridges is greater than the results warrant, and there is the risk of drawing the fire of the enemy's artillery. The guns will work best by sections, and the greatest freedom of initiative is to be left to the section commanders. In action each section has a cyclist to be used to keep in touch.

In the attack the section must push forward with the infantry, but not necessarily close beside it. In changing position the commands are "Démontez"—"En tirailleurs"—"Marche" ("Dismount the guns"—"Extend the section"—"March"). The N.C.O. commands the gun, carries two cartridge-belts, and two of his men carry the gun and its tripod.

It is specially pointed out that the machine gun can never take the place of artillery. There are notes on the action of machine-gun sections on the defensive, on outposts, and in covering a retreat.

The Hotchkiss gun of the original St. Denis type is the machine gun of the Japanese Army. Very few were available at the beginning of the war in Manchuria, but a considerable number were obtained from France during the war, and after the war the manufacture of the gun was begun in the Japanese arsenals, and the number in use in the army greatly increased. At the manœuvres of 1907 a large number were available, and four were attached to each infantry regiment. At the manœuvres of the following year each regiment had a battery of six machine guns, an increase of two per regiment in the twelve months.

The SPANISH Army has both the Maxim and the Hotchkiss gun. The latter is now the standard weapon. Large numbers of Hotchkisses were purchased from the St. Denis factory in 1908, probably as part of the preparation for the war in Morocco.

It may be noted here that in 1905 Kaid McLean introduced Maxims purchased in England into the Moorish Army as its machine-gun armament. But in the narratives of the French and Spanish military operations in Morocco we do not find any instance of the successful use of machine guns by the enemy. It would appear that McLean had not carried the organization and instruction of his machine-gun sections far enough for the Moors to become really familiar with the working of the gun.

The Maxim was the standard machine gun of the ITALIAN Army until 1908. In that year a Royal Commission appointed to advise upon improvements in the armament of the Italian forces, recommended the adoption of a new type of gun invented by the engineer Perino, and known by his name. The report claimed various advantages for the Perino over the Maxim, but one of the chief motives for its adoption appears to have been the desire to give the army a weapon invented by an Italian engineer and manufactured in Italy. The gun is a modification of the Maxim type, using, besides the recoil, a device to produce a reaction of the explosive gas from the muzzle. This seems at first sight a needless complication, but the Italian Commission claimed that it gave a more rapid action to the gun.

It was also stated that the cost of the gun would be much lower, and it was urged that with the large quantity to be provided for the army, this would enable important economies to be made. The gun is water-cooled like the Maxim, and the cartridges are fed into it by means of metallic belts, like those of the Hotchkiss. The organization proposed by the Commission was to attach the guns to existing units at the rate of one gun for each battalion of infantry or regiment of cavalry—a departure from the now generally accepted principle that machine guns should work in pairs. The guns were all to be carried on pack-saddles, on horses or mules—a reserve of ammunition being carried in the same way. It would appear, however, that the introduction of the Perino gun did not entirely supersede the Maxim, for the Italian newspapers of 1909 note the distribution of 120 Maxim guns to the Alpine troops and cavalry and infantry regiments stationed in the North of Italy.

The Maxim is also the machine gun of the DUTCH, SWEDISH, TURKISH and GREEK armies, and most of the armies of South America. The SWEDISH and NORWEGIAN Armies have a number of Madsen guns for the cavalry. DENMARK uses the Madsen in the army and the Maxim in the fortresses. The PORTUGUESE Army has both Maxims and Hotchkisses. The machine guns of the BULGARIAN Army are mostly of the Schwarzlose type supplied from Austria. The RUMANIAN Army has the Maxim. The DUTCH colonial troops have the Hotchkiss gun. The UNITED STATES Army now has the Maxim gun, which has been adopted after many trials, and is manufactured in that country. In many of the armies there has lately been a tendency to introduce side by side with the heavier weapons lighter machine guns of the Madsen or Lewis type, especially for the cavalry.

CHAPTER IX

THE TACTICS OF THE MACHINE GUN

LET us now endeavour to sum up a tactical doctrine of the gun from what we have learned of its history, its use in war, and the opinions of practical soldiers who have commanded machine guns in action and studied its capacities.

Machine-gun fire is *concentrated* infantry fire. This statement is the general guide to its correct tactical use. But, like all general statements, it must be carefully considered before it is put to practical application.

For machine-gun fire has special characteristics that are entirely its own. It can be concentrated like a jet of bullets on a single oval area, or by the traversing of the gun on its pivot it can bring a sweeping fire to bear over a wide front. Thus the machine gun gives to a small group of men the power of either keeping up a slow deliberate fire or delivering sudden gusts of fire, turning it rapidly on a diversity of targets or directing it upon one narrow space of ground, or again sweeping the front with a rain of bullets that produce the effect so well suggested by the French technical expression, *feu fauchant*—a "mowing-down" fire.

The fact that only a few men are engaged in operating a group of guns, and that each gun is fired from a fixed support, with mechanical control of elevation and direction, gives a further special character to its fire. *There is less scope for the errors introduced into infantry fire by the human element.* Nerves and excitement are to a large extent eliminated. A body of infantry soldiers firing the same number of bullets will include a wide diversity of temperaments. As each man reloads and brings his rifle to the shoulder he will have to take a new aim; and experience shows that there are few men who, in the excitement of battle, fire with anything approaching the steadiness of a fairly good shot on the rifle range. No matter how good the general discipline of the men may be, and no matter what earnest and well-directed efforts their chiefs may make to exert control, the firing tends to become excited, the bullets go high. As the range diminishes and the crisis of the fight approaches this tendency increases in a marked degree.

The machine gun, *because it is a machine*, and because it is aimed by one man, delivers an ideally controlled fire. Various estimates have been made of the comparative effect of infantry and machine-gun fire. Some of the most interesting data are supplied by trials carried out by the United States Government about six years ago. The machine guns used were Vickers-Maxims. One of these guns was put in competition with a body of fifty riflemen. These were all exceptionally good shots, selected from a class at the United States School of Musketry. Further, it must be remembered that they were firing, not under battle conditions, but under range conditions.

Mr. Edward Crossman, who tabulates some of the results in a paper in the *United Service Magazine*,* points out that

> The infantry fire was delivered by cool, trained, highly-effective shots, without the disturbing effect of battle to count against them. In action it is far easier to hold down one or two men to cool, well-directed fire, than it is to hold down a hundred men. If the machine-gun pointer proves temperamentally unfit because of lack of coolness, he may easily be replaced. And so in actual battle the fire of the gun, round for round, would be probably more efficient than the same number of rounds fired by infantry.

We may take it, therefore, that the gun was handicapped by the special skill of the riflemen. Yet it more than held its

* April, 1915, p. 65.

own, the advantage being more marked as the range lengthened, and being especially noticeable when *indirect* fire was used against hidden targets. The superiority would be still more marked if the gun were put in competition with a platoon taken at haphazard from an average battalion. And, again, it would doubtless be still more striking if the firing had to be carried out, not under the ideal conditions of the rifle range, but amid the stress and excitement of the battlefield.

It is quite certain, therefore, that the machine gun has at least the fire power of fifty rifles. It is probable that this estimate might be safely doubled. We arrive thus at the conclusion that the probable fire power of a section of two guns is equivalent to the condensed fire power of two platoons of infantry.

And this condensation is the more remarkable if we take into account the fact that the machine guns require only a front of a few yards, while a hundred rifles deployed in the firing-line in the first stage of an attack may cover nearly an eighth of a mile.

Hence we have another characteristic advantage of the machine gun. It is easier to conceal it from view and to secure for it effective cover against fire.

Again, firing from a fixed support it not only keeps its target and range better than even the best-trained platoon of riflemen, but it has a longer effective range than the rifle fired from the shoulder.

But this is not all. Its fire is more effective. One cannot judge the effect of fire in battle by merely counting up the hits made on paper targets on a rifle range. We have seen that the human element must be kept in mind with reference to the men who are firing. But there is also the moral effect on the men who are being fired at. On the range there is no such factor in the fire effect on the targets. But in battle it is all-important.

As is so often the case in discussing military problems, we

are reminded of Napoleon's saying that in war the moral is to the physical as ten to one. In fire effect, not on targets, but on men, the moral effect is everything. One does not win battles by shooting down or bayoneting every opposing man, nor is a campaign decided by the complete destruction of the enemy in the literal sense of the word. Fire is intended to kill or disable a number of the enemy, and to do this in such a way that those who remain will be "demoralized"—that is, reduced to such a condition that they will no longer be steady disciplined soldiers, but will cease to shoot straight, and be so shaken that they will give way before the final rush with the bayonet.

Now men are less impressed and less shaken by a comparatively heavy loss gradually incurred during a long space of time or over an extended front than they are by even a lighter aggregate loss inflicted on them suddenly, in a few minutes, and on a small space of ground. In the first case they hardly realize their losses and their danger. In the latter they are subjected to an intense moral strain. There is all the difference that exists between the prolonged pressure that a steel bar will sustain almost without bending, and the sharp blow that will shatter it.

The machine gun supplies the means of delivering this sharp blow. Its gust of destructive fire has a peculiarly nerve-shaking quality. Those who have to face it and witness its devastating effect on their comrades have the uncanny feeling that they are up against a machine, not merely fighting with other men. And the effect is all the more demoralizing when the machine itself is invisible and there seems to be no possibility of doing anything to put it out of action. To this we must attribute the well-recognized fact noted in so many accounts of the action of the guns in battle, namely, that men seldom fail to remark the peculiar rattling reports of the machine guns (at least four different timings in reports—two German and two British), and are heartened and encouraged by hearing it on their own side, and depressed by recognizing it as it dominates the crackle of rifle fire from the attacking line, which they are trying to hold back. The machine gun has thus some share of the moral effect that belongs to artillery in action.

The machine gun, properly mounted and in the hands of duly trained men, should be as mobile as infantry in the actual fire fight. Modern machine guns have all been improved in the direction of lightening both the gun and its mounting. In moving from position to position in action the gun is light enough to be carried or dragged along the ground by one or two men. The guns are far more mobile than artillery, and compared to the field gun, with its wheeled carriage and its teams of horses, the machine gun presents an infinitesimal target, and of such small height that it can generally be moved under cover.

It has already been noted that in range tests against targets hidden by an intervening obstacle machine-gun fire gives better results than that of infantry. This point is worth insisting upon. If the distance and direction of the unseen target is known, and it is within reach of even the longest range of the gun, the target can certainly be hit if the trajectory of the cone of bullets clears the intervening obstacle. And when one says it can be hit, this means that it can be kept under a steady rain of bullets. *It is not a matter of chance but of certainty.* With correct elevation, direction, and the gun fixed on its mounting, the descending cone of bullets will fall upon the same patch of beaten ground as long as the gun is kept in action.

These are the advantages of the machine gun. It supplies "condensed infantry fire," and this condensed fire has characteristics that make it more effective than the fire of the steadiest infantry, if the machine gunner knows how to develop to the utmost the powers of his weapon.

Rate of Fire

Every gun that can be fired rapidly has a certain drawback, the importance of which used to be greatly exaggerated. In the early days of machine guns it was objected that the gun would consume ammunition at a tremendous and even a prohibitive rate, and would be liable to be put out of action by the mere want of cartridges. But any such objection is now of less force. The adoption of quick-firing artillery and magazine rifles was opposed on the same grounds by the same conservative-minded critics who find objections to every progressive change. In the case of machine guns the objection was largely based on a misconception. The objectors imagined that in battle the guns would be in action, and actually firing, for hours at a time. Experience shows that a machine gun properly handled is at work in most cases for a few minutes at a time. Fire is used when it will tell and tell heavily. The machine gun is fired to hit something, and is not to be handled as a mere cartridge destroyer. Nevertheless the aggregate consumption of cartridges may be very high, and the supply must always be ample. But this is a matter of starting with a good supply in hand and having proper arrangements for replenishing it. *It has been truly said that the winning of battles now depends to a great extent on the ammunition supply being previously arranged in a methodical manner with a definite view to the intended operations. This is eminently true of the machine-gun arm.*

Another drawback has also been made the subject of exaggerated criticism. In the early days of machine guns a distinguished officer of our own Army said that he found that the gun had an awkward trick of getting out of order just at the precise moment when it was most wanted. This was largely at the time the result of the defective cartridge then in use, the old Boxer pattern with its composite case, which was liable to go to pieces under the pull of the extractor, thus blocking the loading chamber and jamming the mechanism. The solid-drawn cartridge-case has put an end to this source of trouble. But it is true that machinery of any kind is liable to unexpected stoppages.* The mechanism of a machine gun is not much more complicated than that of the modern magazine rifle. But in the case of the gun it works at a high rate of speed and is subjected to a rapid succession of shocks, several hundred in a minute. The wonder is that breakdowns and jams are so rare. When they do occur, in ninety-nine cases out of a hundred they can be set right within twenty-five seconds. But the fact that a stoppage can suddenly occur is the basis of the sound tactical rule that machine guns must always work in couples. Two guns form the smallest tactical unit, and we have seen that two guns represent a very considerable fire power.

The drawbacks enumerated are therefore met by a carefully organized ammunition supply and the working of the guns in sections of two each. Two other points may be noted. Machine guns, like artillery, are for the time being out of action while on the move, though they have the advantage that in *well-trained hands they can come into action instantly.* The practical deduction is that once the guns are in a good fire position *they should not be moved without reason;* the movements should be as rapid as possible, and the new positions should be selected before the guns are moved.

Again, the machine gun has no place in the fight at close quarters. It follows, therefore, that during the attack it keeps in action as long as its fire can be maintained without danger to the assaulting troops. When the fire has to cease the guns are got ready to push forward and assist in holding the captured position against a counter-attack. On the defensive in the same way, the guns will be kept in action till the last moment against the advancing enemy, but during the actual

* With thoroughly trained gunners stoppages will be rare and only momentary for the most part.

struggle for the position, and when the attacking force is penetrating into it, the machine guns should be withdrawn and held ready, either to cover the retirement or to be pushed forward to reopen fire on the retiring enemy, if the assault is repulsed.

Machine-gun tactics have suffered somewhat from the false analogy between machine guns and artillery. Some trace of this false view still lingers, and there is evidence of it in the fact that during the present war captures of machine guns on both sides are enumerated as if they were captures of cannon. The Service Regulations of all armies now lay it down that there are occasions when guns can be honourably lost in battle. As, for instance, when a battery sacrifices itself to cover a retirement. This is true, even in a greater degree, of the machine gun. The weapon is so easily manufactured, *and should be available in such large numbers*, that the loss of a few guns is not really a serious matter. *It is much more difficult to replace the trained officers and gunners of a section than the guns themselves.* Too great anxiety to prevent the guns becoming war trophies for the enemy may easily lead to an absence of enterprise in the attack, and on the defensive it may have an equally unfortunate result. There is the danger of guns prematurely ceasing firing in order to be withdrawn, perhaps ceasing fire at the very moment when their rain of bullets was shaking the near advance of the enemy. The machine gunner must be prepared to risk the loss of his gun, and prepared also to disable it at the last moment, so that it shall be of no use to its captors.

A practical working system of machine-gun tactics must be based upon the effort to make the most of the characteristic powers of the weapon. If it is to be used effectively, it must be in the hands of officers and men thoroughly familiar with their guns and imbued with the enterprising spirit that will seize and make the utmost use of every occasion for their intervention in the fight.

In the various opinions on machine-gun tactics which are quoted or summarized in the preceding chapter, it is evident that the writers had in view the defensive attack of an enemy's position in pitched battles fought in the open. In the present war we have been brought face to face with a state of things which is not really new in its essential features, but is new in the extent of ground over which a special condition has prevailed and the length of time during which this condition has dominated the operations, especially on the Western front. Instead of battles in the open and a war of movement and manœuvre, we have what are virtually prolonged siege operations between armies entrenched on prolonged fronts and close up to each other. *We have said it is not new, because siege warfare is as old as war itself, and, further, the fight for entrenched positions with both sides dug into the ground has been a feature of many recent wars, beginning with the Wilderness campaigns in the American Civil War.* The war of entrenchments has played a part in every campaign since Plevna. The battles of Liao-yang and Mukden in the Russo-Japanese War were prolonged entrenchment fights. But in this war the entrenched fronts in the Western sphere of operations are of a *length* for which there is no precedent in any previous war, and the condition of stalemate has lasted for months.

Manœuvre Battles

Those who are training our machine gunners have, therefore, had their attention riveted on the question of their employment in this prolonged entrenchment fighting—a very different business from their use in manœuvre battles, and one hardly contemplated by early writers on the subject. In conversations with machine-gun instructors, one finds indeed a tendency to leave entirely aside the question of machine-gun tactics in the open. But, as the records of the war in Eastern Europe show, there is still a place for the *manœuvre battle*, and the war of entrenchments is not everything. But the conditions of these two kinds of fighting are very different, and in discussing the tactics of the machine gun there must be a separate treatment of the machine gun in the manœuvre battle (which has been so fully discussed in recent years) and the use of the machine guns in the effort to break an entrenched front by an operation which on a vastly larger scale is similar to the old assault on a besieged fortress.

Effective Range

So far as the machine gun is concerned, a radical difference between the two kinds of fighting arises from the question of the range at which the guns are brought into action. We have seen that in the evolution of machine-gun tactics there has been a steady tendency to regard medium and short ranges as those at which the gun can be used with best effect in battles in the open. This arises from the fact that normal machine-gun fire is concentrated fire with a very limited beaten zone—*a fact which suggested the rule that fire, of which the effect could not be observed, would be wasted.* It was felt that the efforts to bring long-range fire to bear upon an enemy's supports by spraying the ground behind his firing line with machine-gun fire, would generally be mere random work depending on pure luck for serious effect, and that in most cases it would mean a costly and almost useless expenditure of cartridges. This was held to be work for the artillery, not for the machine guns, and that guns supporting infantry could be of the most service to them by holding their fire until medium ranges were reached and then using it as the opportunity offered to support the advance.

The same considerations led most writers on machine-gun tactics to regard indirect fire as something exceptional, peculiarly difficult, and of doubtful effect. The tendency therefore was to concentrate attention chiefly on the *use of machine guns with direct fire and at ranges of a thousand yards and under.*

But the fact remains that while infantry fire at long ranges is only exceptionally effective, the machine gun on its fixed mountings can range up to 2,800 yards, and in properly trained hands can make good shooting at this extremely long range. Further, the height of the trajectory at long ranges favours indirect and overhead fire by making it a simple matter to clear intervening obstacles and to fire safely over the heads of one's own advanced troops. In view of these facts, one asks oneself if with a gun ranging up to 2,800 yards there is no means of making efficient use of its powers beyond less than half this distance. Experience of machine-gun work in the entrenchment battles of the Western front shows that, whatever may have been written on the subject before the war, *there are plenty of means of using the guns at even the longest ranges in this new kind of battle.* It remains true that machine guns must not be confounded with artillery, but nevertheless some of the recent developments in their use tend to assimilate it to those of the heavier weapon.

In most of our training camps at the present moment it is laid down as a general rule that the effective ranges of the machine gun lie between 1,000 and 500 yards. This, however, refers to machine guns acting with infantry and pushing forward with them in the advance. The use of the guns at longer ranges is for covering fire over the heads of the infantry they are supporting, *or directed against the enemy's supports* and the lines by which he is bringing up his reinforcements to the fighting-line. This long-range fire will often be indirect against targets unseen by the gunners themselves, and on account of its high angle it will generally be somewhat similar to indirect fire. We have seen that the view which for a long time barred the use of long-range indirect fire for machine guns was based on the quite sufficient reason that in most cases the results obtained would be doubtful and out of all proportion to the amount of ammunition expended. But this objection is removed by methods more recently worked out and rendered possible by the special conditions of the prolonged entrenched battles of to-day. We may say, indeed, that there are now two kinds of machine-gun tactics: *the tactics of long range*, rendered possible by the conditions of the entrenchment battle, and *the tactics of medium and short ranges*, which have their place in the manœuvre battle in the open, and the assault during the entrenchment battle.

The special conditions that render long-range fire practicable and effective are these. In the entrenchment battle prolonged not only over days, but it may be over weeks, the enemy's position is fixed and easily defined. More than this, systematic aerial reconnaissance, day after day, renders it possible, not merely to fix the general position and limits the hostile position, *but also to map out most accurately the position of the advanced trenches which form his firing-line, the trenches farther back, where he keeps his supports and reserves, and the lines by which these supports and all supplies of ammunition must be brought up to the advanced trench, these lines being the communication trenches.* Not only is the enemy's firing-line permanently fixed to a defined position, but all movements immediately in rear of it must necessarily follow clearly fixed lines.

During the long preparation for an attack upon the enemy, all these positions and lines of communication can be accurately laid down, if large-scale maps of the ground he holds are available. In the warfare on the Western front these maps, elaborately contoured at short vertical intervals, are available, and this not only facilitates the mapping of the enemy's position, but also makes it perfectly easy to work out rapidly accurate sections of the ground on any line of fire in its front. It is therefore possible to select a machine-gun position in our own lines, or in rear of them, from which by indirect fire, selecting the appropriate range and trajectory, the bullet sheaf from the guns will clear the intervening obstacles and descend upon a given spot in the enemy's lines. *That the bullets will strike the selected patch of ground in the enemy's position is not a matter of chance, but of absolute certainty.* This would not be the case with long-range fire from a platoon of riflemen. A number of men, no matter how well trained, will not go on steadily bringing the rifle again and again to the shoulder with absolutely the same elevation and direction. But the machine gun, once clamped on its tripod, delivers its fire with mechanical certainty on the same spot.

It is quite true that, even with the help of aeroplanes, there cannot be the same easy observation of fire effect as in the case of artillery with its bursting shells. There can, however, in many cases be observation of fire, either from the air or from the ground, but this is not necessary to make this kind of long-range fire effective. The important point is that reconnaissance of the enemy's position can reveal the *areas where it is vulnerable to machine-gun fire*, and this fire can be directed on these areas with certainty, and with the result that the fire of the guns will harass the enemy, impede his movements, and inflict loss upon him.

A few details will make the matter clearer. At these long ranges the fire of the machine gun is no longer a closely concentrated jet of bullets, but spreads over a beaten zone, oval in general form, somewhat like the beaten zone of shrapnel, but not so large; its major axis will be in the direction of the line of fire. With lighter guns, like the Colt, the beaten zone will be rather larger than with the heavier and steadier Maxim. In this case the slight vibration of the Colt and similar guns is an advantage rather than a drawback, for it tends to enlarge the beaten zone, whilst still keeping it within sufficiently moderate dimensions for accurate fire.

To take some practical instances, the enemy's communication trenches immediately behind his first line have been located and mapped. During an assault it is certain that he will be sending up reinforcements and supplies of ammunition along these trenches. Positions are therefore selected from which machine-gun fire is brought to bear upon each of those communication trenches. In selecting these positions, sections are drawn of the intervening ground in order to select the range and trajectory that will be most effective. It is seldom that fighting takes place on perfectly even and level ground. Even what we call plain is in Western Europe, in most cases, a series of gentle undulations. Reserve trenches will often be placed in rear of the crest of a slope of ground. In such a case one will find the communication trenches running back by the reverse slope of the rise in the ground to support positions in rear of and hidden by it. In such a case it will often be possible to select a machine-gun position that will give a trajectory which will sweep the reverse slope. The angle of descent of the bullet sheaf conforming very closely to the slope of the ground, such a fire will be very deadly and embarrassing to the enemy; it may even prove to be a barrier through which he will find it impossible to push forward his reinforcements.

It is obvious that the longer the range the more chance there is of selecting lines of trench that are open to enfilade. In the entrenchment battle the advanced trenches on both sides are fairly near each other at the outset. By using the longer ranges of his weapon, the machine gunner with his guns in position behind the advanced lines can pick out lines of trench far away to right or left, on which he can bring a diagonal indirect fire, sometimes enfilading a considerable length of trench. It is all a question of having absolute confidence in the gun, bringing its rain of bullets to bear on the selected, though perhaps unseen, point, and a careful preliminary study of the forms of the ground and the possible trajectories of the weapon.

The beaten zone of machine-gun fire, even at long ranges,

is so moderate in extent that several guns will have to be used together to sweep a given extent of ground either in frontage or depth, though the single section of two or four guns is quite sufficient when firing upon a narrow target, such as a communication trench running up to the enemy's front. The number of guns to be used will be settled by the extent of the target. The range being given, the machine-gun officer can say, "With so many guns at such a distance, I can cover such-and-such a space of ground with a steady rain of bullets." Actual experience has shown that he has thus the power of inflicting serious loss on the enemy's supports, and even driving them from point to point in search of cover or escape from the showers of bullets that are descending from some unseen firing-point. *Colonel Mayne years ago* (1903) *argued for the possibility of sweeping with rifle fire the reverse slope of hills or rising ground behind which an enemy was sheltering his supports and reserves.* The idea was perfectly sound in principle, but a difficulty arises in its realization in practice, from the unsteadiness and uncertainty of indirect rifle fire. The steady certainty of the machine gun remedies this, and makes Colonel Mayne's idea of indirect fire with trajectories conforming to the slope of the ground perfectly practicable.

One need hardly point out that, with gunners trained to use long-range fire in the way which has been described, it will be a simple matter for them to use the same kind of fire over the heads of their own infantry in support of an assault. Many writers on tactics show a reluctance to admit of the firing of machine guns over the heads of advancing infantry and in support of it, on account of the danger of the gunners inflicting loss on their own friends. With the least care as to watching the advance of the infantry, this danger utterly vanishes with the high trajectories of long-range fire. In fact, it can only exist when the guns are firing from positions close behind the infantry, and therefore at medium and short ranges.

Siege Battles

There is, of course, in the entrenchment battle, which is a prolonged series of trench operations, and in the assault, which is its culmination, abundant scope for the use of machine guns at short ranges, as well as for this elaborately prepared indirect and high-trajectory fire at long ranges.

The use of machine guns at short range in the assault approximates to its use in the manœuvre battle in the open. But there are certain differences. In many ways the task of the machine gunner is simpler. Thus, there is no question of the gradual advance from position to position during the preliminary fire-fight of the battle in the open, an advance which requires some of the highest qualities of leadership. *In the entrenchment battle* the position for the guns at the outset will have been deliberately chosen beforehand. There will be selected points in the advanced line of trenches to which the guns will, of course, be brought up under cover by the communication trenches, probably during the night. There will be no difficulty about ammunition supply, for an abundance of cartridges will have been collected at the gun positions. The enemy's line having been carefully mapped in advance, the choice of targets will be easy and the ranges can be fixed with absolute accuracy. At the outset these will be very short and all within fixed-sight range.

There is a special kind of preparatory work which is sometimes assigned to machine guns in the advanced line before the assault. They may help in the destruction of the wire entanglements in front of the enemy's works. They cannot do much damage to the wire itself, but the supporting posts can be cut down by the stream of bullets from the gun. It is an application in practical warfare of the trick with which Sir Hiram Maxim used to impress the spectators at trials of his gun when he was first exhibiting it. He used to cut down a tree three or four hundred yards off by firing upon it with a short traversing motion of the gun to distribute the bullets across its entire width. In this way he often brought down a large tree with a single belt of cartridges. In firing on an entanglement far fewer shots are necessary to bring down a supporting post, for once it is badly damaged the tension of the wire support will often be quite sufficient to break it off near the ground.

Once the assault is launched the guns in the advanced trenches will push forward after the first wave of the attack. Having got into the enemy's position, their work will be to assist the infantry in clearing points to which he still clings, breaking up any attempt of the defence to rally or to counter-attack.

In a close fight of this kind there are greater chances of enfilading communication trenches and trenches of the second line, and in the fight for villages within the enemy's line the guns will be able to use their concentrated fire in clearing the streets.

It is held by some that machine guns should not attempt to engage the machine guns of the enemy, leaving to the artillery the work of silencing them. But in village and trench fights there may be exceptions to this rule. There will often be a chance of opening fire on hostile machine guns, not directly from the front, but obliquely or from a flank, and the more suddenly the fire is opened the more effective it will be. In village fighting guns are sometimes found mounted inside a window on the ground floor or the first story of a house. It is difficult to silence them with infantry fire, but we have here another case where a machine gun may be usefully brought into action, not, of course, in the open and under the direct fire of the enemy's machine gunners, but after bringing it up under cover behind a wall or other obstacle, or inside of a house, from which the hostile machine-gun posts can be seen. The action of the gun at short range is then like that of a fire-hose directed upon the window that is attacked.

In the fighting inside the village of Loos on September 25 it is said that effective use of machine guns was made in clearing the houses by bringing them up to close quarters and firing a few rounds upon them into the lower windows of the houses. The rapid reports of the gun were sufficient to stampede the enemy holding the houses attacked.

The lighter type of machine gun, without a belt feed, such as the Lewis, Hotchkiss and Madsen, are the best adapted for this kind of close fighting, which will often entail hurried movements under cramped conditions.

But perhaps the most important work of the machine gun in the assault will be the part they will take in securing permanent possession of the captured ground. Attempts at counter-attack will be practically certain to be made, and it will often happen that the direction of these attacks can be easily foreseen, and the lines of approach for the enemy will be restricted by existing obstacles, and the tendency to use communication trenches and similar partially covered lines of advance. The attacking force will also tend to collect in a mass of men either from the very outset or at the first check to those in front. The range will also be short. Also help from long-range machine guns. All these conditions are favourable to machine-gun effect, and it will be essential to have a number of guns got into position in the captured works at the earliest possible moment, and placed so as to sweep the probable lines of counter-attack. A captured line of works will in most cases have an irregular frontage, some portions of it lying well to the front of the main line, and these advanced points will afford good positions for bringing the enfilade fire of machine guns to bear on the counter-attack, and at the same time such a salient position can be strengthened by placing guns farther back to right and left of it to sweep the fronts and cross their fire beyond the angle of the salient. These points suggest that it will be useful to have an officer (Divisional M.G. Officer) detailed to direct the general organization of the machine-gun defence in the captured position, instead of leaving it to officers commanding smaller groups of guns to push them forward here and there in a somewhat haphazard way, suggested by their own impressions of local needs and *without taking into account the necessity of co-operation, which will make the fire of a smaller number of guns thus organized on a general plan even more effective than that of a far larger number put into the front here and there by individual section commanders.*

The tactics of machine guns on the defensive in entrenchment warfare hardly need any lengthened explanation. What has been said of the use of the guns in securing a captured position applies equally to their use in protecting the advanced lines of entrenchments against a possible assault or some local enterprise of the enemy. Entrenched positions are now held for such a length of time that their defence can be as elaborately arranged as that of a permanent fortress.* With a comparatively small number of machine guns installed in carefully concealed and thoroughly well-protected positions, provision can be made for sweeping with flanking fire long fronts of the

* In their arrangements for defending their entrenched positions, the practice of the Germans seems to be to have very few guns in the first-line trenches, but considerable numbers of them in the second line. This arrangement is based on the theory—justified by the experience of many attacks made from both sides—that, given a sufficiently powerful artillery preparation, the first-line trench can be nearly demolished, and then easily rushed along a considerable front. The further progress of the assault then becomes much more difficult, and the really serious fighting is for the second line of trenches, and the machine guns massed there are a most valuable element in their defence in this critical stage of the battle. In several narratives of the battle of Loos one finds officers and men who took part in the assault telling of the heavy loss inflicted by machine-gun fire in the advance beyond the first line of trenches—losses far exceeding anything that was incurred in the rapid storming of this first line.

advanced trenches, and at the same time guns can be used from positions farther back to bring a high-angle fire to bear upon the supports of a hostile attack. The mobility of machine-gun sections and motor machine-gun batteries also provides a ready means of reinforcing any point of the front that is attacked.

NIGHT-FIRING

Various methods have been devised for using machine guns at night in repelling attacks on an entrenched position. Many of these plans work out fairly well on the rifle range, but in actual fighting it may be said that night-firing is only effective with machine guns when the simplest arrangements are adopted. Complicated devices are apt to give disappointing results. Where the target can be illuminated with the help of searchlights, star-shells, flares, and the like, the conditions of the firing approximate to those of daylight. But where the firing has to be done in the dark, without such helps, it is safer to be content to use it only where the guns can be laid in daylight on a limited line of approach over which the enemy must come—a road, a bridge, a defile and the like; to clamp the gun in the required position; to keep to the fixed elevation and use only a very limited amount of traversing. There are various plans for marking during daylight the various positions in which the gun can be placed, and then putting it into position and opening fire in the dark, but all these complications give many openings for error, and the only safe rule is to use none but the simplest methods.

MANŒUVRE BATTLES

Before the war writers on machine-gun tactics devoted their attention chiefly, or even exclusively, to the use of the guns in the manœuvre battle, fought over open ground. But the prolonged fighting on entrenched fronts, which has been the predominating feature of the war in the West of Europe, has tended to concentrate attention almost entirely upon this curious combination of battle and siege methods, but the tactics of machine guns in the open have still their importance. It is obvious that, where armies have become committed to the prolonged warfare of entrenchments, the result is a kind of stalemate or deadlock, and that a decisive result can be obtained only by one side or the other breaking through the opposing entrenched lines, this success being the prelude to a series of further movements that will involve manœuvre battles.

It is true that even in these battles there will still be entrenched positions, for nowadays no body of troops will halt even for a single day without strengthening its position and digging itself into the ground; but the entrenchments, on account of lack of time, will not be of a very formidable character. They will be very different from the elaborate and complicated works that have gradually grown up during long months along the opposing fronts in France and Flanders. They will be more like the shelter trenches, gunpits, and other light field-works, which we find described in the drill-books of many years ago (1893), and on both sides the main resource for securing cover will be the use of the existing natural features of the ground. Under these conditions we shall have again battles fought out in the open.

In such a battle the most important work for the machine guns will be to support the infantry attack. This support can be given in two ways—by long-range covering fire, and by the fire of machine-gun sections accompanying the infantry advance.

As for the first method. There certainly will be occasions when the long-range fire of the machine gun may be usefully employed. In discussing machine-gun tactics in the entrenchment battle we have described methods by which this kind of fire has been effectually used at ranges up to 2,800 yards. The special conditions of the entrenchment battle, the thorough mapping of the ground, and the accumulated results of aerial reconnaissance, make this kind of fire easier to employ than it will be in battles in the open. But this only means that in the latter case it will not be so frequently employed as in the former. The experience of entrenchment warfare has shown what are the powers of the weapon at long range, and it would be irrational to lay down fixed rules that no attempt should ever be made to use the machine gun under somewhat analogous conditions in the open, or to fix a thousand yards as the extreme limit of its useful employment.

We would suggest that there is all the more reason for endeavouring to develop the use of the gun at long ranges, because if it can be successfully employed in this way it would meet a practical requirement of the modern battlefield. A few years ago the normal range at which field artillery came into action was frequently fixed at about 2,500 yards as an ideally satisfactory distance. It was after the experiences of the war in South Africa that field artillery ranges were lengthened. Until then our shrapnel fuses were designed for action at ranges which the artillerists of to-day count as medium or short, not long. With the lengthening of the artillery range we have left between the normal artillery positions and the ranges at which infantry come into action an intermediate zone, for which there is no approximate weapon. The long-range fire of infantry is notoriously of problematic value, and effective only under very special conditions, with very highly trained men and with exceptional leadership. Our infantry rifle is sighted up to more than 2,500 yards, but in practice is hardly ever fired at even half this distance. At ranges above a thousand yards machine-gun fire has been proved to be far more accurate than infantry fire, for reasons that have already been explained.

But if we take the broad zone of possible ranges from 1,000 up to say 2,500 yards, we will find that, as a rule, artillery will not be brought into action at these ranges. It will generally be easier to find new positions for it beyond the 2,500 yards limit. At these longer ranges its fire will be effective, it will have a larger choice of targets, it will be easier to concentrate the fire of several sections on a given point, and there will be greater facilities for cover and concealment from view.

On the other hand, machine-gun sections can find positions which will give them sufficient and even ideal cover on grounds where it would be impossible to find shelter for a battery of artillery, and we would suggest that the normal ground on which the machine gunners should seek for positions for long-range covering fire in support of the infantry attack should be in this intermediate zone between the long-range infantry positions and the nearer artillery positions, a zone of ground nearly a mile wide. The machine guns would thus fill a gap in the arrangement of the various zones of fire on the battlefield. Their action will be analogous to that of the long-range machine guns in the entrenchment battle, which has already been described. The section of machine guns will probably not have to change its position during the fight. It can somewhat deliberately select the ground on which it will come into action, the chief requirements of which will be good cover, which can be artificially improved, and a view to the front that will enable fire to be brought to bear upon the point attacked; or, alternatively, indirect fire can be used with the help of an observation station, and the gunners need not have a direct view of their targets.

More immediate support can be given to the infantry attack by the machine-gun sections which actually accompany it. Before the present war it was a generally accepted idea that machine guns could find a place even in the actual firing-line, but one sees now the tendency in some quarters to regard this as impracticable, and restrict their activity on the battlefield to covering fire. Those who hold this view suggest that if it is desired to increase the volume of fire in the actual front of the attack by mechanical means, this can be done by arming a certain number of the men in the firing-line with automatic rifles, or if machine guns are used for this purpose, giving them guns of the type that approximates to the automatic rifle, such as the Lewis gun.

There is a possible objection to this suggestion, arising from the question of ammunition supply. Whatever kind of rifle the soldier carries, whether the ordinary magazine rifle or the automatic, or the Lewis gun, the increase of fire power can only be secured by his having a supply of ammunition greater than that which the infantry soldier ordinarily carries. The limit of this possible supply is soon reached, and if it is to be increased the soldier must be accompanied by ammunition-carriers. We thus come back to something very like a small machine-gun section. If we are to have men in the firing-line doing the actual firing, while others attend them as ammunition-carriers, we have the group of men which the opponents of the employment of machine guns in the firing-line regard as a dangerous target to hostile fire. The objection is not serious, for the group can work together without actually standing side by side and attracting attention. But once we recognize the possibility of placing such a group in the firing-line, there is no reason why the group should not be provided with the more powerful guns of the Maxim, Vickers, Colt, or Hotchkiss type, and have the advantage of a second group with a reserve of cartridges working under cover somewhere in their immediate rear. This gives us the machine-gun section, and the experience of actual work on the battlefield in earlier wars shows that, with good leadership and proper use of ground, such a section of machine guns can find a place in the firing-line without necessarily drawing upon itself a destructive hostile fire.

We take it, therefore, that in coming battles in the open, besides the long-range covering fire of the machine gun in what we have described as the intermediate zone, it will be well to have machine-gun companies accompanying and working with the infantry in the attack, in order to increase their fire power and give them the advantage of suddenly concentrating an intensely powerful fire upon a given area in the hostile line when the opportunity offers.

The conduct of such machine-gun companies will require very high qualities in their commanders. They will not have an easy task to perform. They will have to be complete masters of their weapon, have a good eye for ground, a sound and ready tactical judgment of the situation, and the enterprising spirit that will seize and make the utmost use of every occasion for their intervention in the fight. These opportunities are often very fleeting. The chance of using the guns presents itself, and if it is not recognized and seized at once the situation changes and the occasion has passed away.

It is obvious, therefore, that a considerable independence of action must be left to the machine-gun company commanders, and their own initiative must be largely relied upon for their timely and effective co-operation with the troops of other arms with which they are acting. It has long been recognized that on the modern battlefield, once an attack is launched, the direction of the actual fighting necessarily passes into the hands of officers commanding small units. In the infantry attack the battalion commander has to content himself with giving general directions at the outset to his company officers, and relying upon them to act as circumstances may require once the advance begins. It follows, therefore, that once the machine guns are sent forward the same liberty of action must be left to the commander of a section of guns that has already been conceded to company and platoon commanders. In the case of the machine guns this is all the more necessary because their tactical handling must depend entirely upon the judgment of the officer who commands them, and the opportunities for their effective action has to be seized at once.

The machine-gun officers will therefore be given at the outset of the attack complete instructions as to its general plan, and a direction as to the way in which they are expected to co-operate. After this all details of execution must be left entirely to them. *If the machine gunner cannot be trusted to make the best use of his guns, he should not be in command of them, and the only man who can judge as to what is possible or useful is the man behind the guns.*

Even if machine guns are organized in companies, the section will still be the tactical unit. Where the guns are all together the company commander can direct them to a certain extent, but he will direct them in the best way if he handles his sections as the battalion commander handles his companies—namely, by giving only general instructions and leaving to the section commanders as great a freedom of action as possible. This is not asking too much, for we have seen that there is good authority for regarding the fire of a machine-gun section as practically equivalent to that of a platoon of infantry.

Our Manual of Infantry Training lays it down that there can be no set form for the attack. General principles are stated, but the circumstances of each occasion are the only guide as to how these are to be carried into effect. When the machine-gun sections become part of the attacking force the same rule obviously holds good. One cannot fix in advance what their place is to be in any and every attack. All that one can do is to note some general principles to guide the machine-gun officer.

In the chapter on the evolution of machine-gun tactics we have seen that two different opinions have been put forward as to the place of the guns in the opening stage of the battle. According to some writers, they should be held in reserve at the outset, until the opportunity comes for pushing them into the fight. According to others, they should be with the infantry in the firing-line from the very beginning of the advance. There is something to be said for this latter view. Amongst the considerations which recommend it one may note that in the first stage of the attack there will occasionally be opportunities for taking advantage of the long-range fire of the machine guns, and further it is obvious that there will be many situations in which it will be easier to carry the guns forward under cover if they start with the firing-line than to bring them up to it when it has already made some progress and its movements are being carefully watched by the enemy's observers. The movement of the guns into the line may in such cases betray their position at once and expose them to the chief danger that machine guns have to face—namely, the deliberately concentrated fire of hostile artillery.

It is, however, possible to take a middle course *now that machine guns are being multiplied in numbers in all armies.* The question of putting them into action at once or holding them in reserve was more difficult to solve when only a few guns were available. *Now that there are many of them, one may perhaps take it as a good working rule to divide the force, put some machine guns into action at once, and hold the rest in reserve for a while, to be sent up as the firing-line is reinforced, or to be pushed forward to bring a storm of concentrated fire to bear as ordered.* A reserve of this kind is so valuable that it is well worth while to keep it in hand until the development of the fight shows where it can best be used.

It may perhaps be objected that this is the revival, in the case of the machine gun, of a tactical theory that was tried and abandoned in the case of the artillery. It was long the fashion to divide the artillery into the field batteries that were put into line at the outset of a battle and the reserve of artillery that was held in hand to be pushed in at a critical moment. The accepted doctrine now is that, unless under exceptional circumstances, to keep guns waiting in reserve is to sacrifice fire power. But in the case of the machine guns we must guard against being misled by the old false analogy between the field gun and the machine gun. The rule always holds good that machine guns are not artillery, but are "condensed infantry fire." *The reserve of machine guns do not represent fire-power left idle, but they should be classed with the infantry supports and reserves kept in hand to be used to reinforce and carry forward the firing-line.* It will only be under exceptional circumstances and with small forces that it will be advisable or even possible to put every rifle into the fighting-line at once, and the same holds true of the machine guns.

The final orders for the attack will, if possible, be issued with the actual ground in view. With the machine guns as with the infantry, it is all-important that a good start should be made. It is almost impossible to remedy at a later stage any serious error made in the launching of the attack. In its preparation there should therefore be an absence of hurry. This does not mean that time should be lost, *but the worst loss of time arises from mere restless precipitation. If the machine-gun officers are properly trained, they should be able to carry through the deliberate preparations for the opening of the attack very quickly. It is important to have a good look at the ground, and previous practice in the study and appreciation of ground will make the work of reconnaissance a matter of a few minutes.* The point to be kept in view is to look out for a first-fire position for the guns, and the best way of getting them up to it, *and then with the help of the map and such view of the ground as can be commanded to try to get a general idea of where further positions are to be found as the firing-line presses forward.*

Obvious facts are sometimes forgotten, therefore it would be well to note that, though riding on to the actual ground and remaining in the saddle gives an officer a better view and enables him to move about more rapidly, it is a sound rule to dismount under cover and do the actual reconnoitring on foot. Even then one must take care not to stand out in the open, and the range-finder who accompanies the officer should not stand close to him, and should be equally cautious about being seen from the front. *Keeping low down and under cover, and looking to the front from a kneeling position, has the further gain that the reconnoitring officer will get the same view of the ground that will present itself to the machine gunners when the guns are brought up.* This careful use of cover in the reconnaissance is all-important. Even at long range there are hostile eyes on the watch, with good field-glasses and telescopes to help them, and the appearance of an officer and a range-finder moving about and examining the ground will suggest to the enemy, if he sees them, that perhaps not machine guns, but even field guns will soon make their appearance, and the enemy's gunners may then be on the look-out to pour a storm of shell on the very position that has been selected.

Co-operation with Infantry

We suppose that the training of the machine-gun section has been carried out in a business-like way, and therefore that in peace time they have had opportunities of manœuvring with infantry in large areas of various kinds of country. Now that they are about to take their place in the fighting-line, working beside the infantry soldiers, the place they will take should be decided by the fact that if they are to be of any use *they must be allowed to place their guns on the ground that suits them best.* The infantryman must advance directly to his front, and cannot diverge to right or left in search of cover. He must make the best use he can of the ground on which he happens to find himself, but the machine guns absolutely require cover of some kind. To bring them into action in the open will be to draw fire against which they have no chance of escaping being put out of action. In the firing-line, therefore, the machine gunner has a right to the choice of ground, and a section of machine guns requires so small a space and so limited a front that this is not a great concession to ask from the infantry it is supporting.

The necessary conditions of a machine-gun position are: a good field of fire; cover, both from view and from fire, and if possible a secondary covered position within easy reach for the ammunition supply. It is a gain if there is water near at hand, otherwise a reserve supply must be brought up with the guns.

As to the field of fire, the essential point is that there should be an unobstructed view of the points on which the attack is first being directed. It is a further gain if there is clear ground for fire to the immediate front of the guns, in case of having to meet a counter-attack. With guns covering the flank of an advance, a field of fire to the right or left is also important. With clear ground in front of them the guns need no escort, but can protect themselves.

As to cover, it will very frequently be possible to find sufficient natural cover for such a small object as the machine gun and its gunners. In the attack there will often be the opportunity for improving this ready-made cover by doing a little work with spade and pick, or placing a few sandbags in position. But it will be rarely that anything so elaborate can be attempted as would be done upon the defensive.

In battle it will be rarely that the attack will be pushed forward at the speed of a field-day advance. *Once the serious fight begins, the advance is usually slow. The guns must go forward with the infantry. But this does not mean that they will move with every rush of the troops beside which they first found themselves, or that they will always be on the same alignment as the neighbouring infantry. It will frequently happen that they can, without immediately moving, support by their fire for some time the troops that are pushed forward to right and left of them. Once the guns are in a good fire position they should not be moved without good reason.* While moving they are out of action, and a good firing position cannot be found at every few yards. When the move has to be made, the next halting-place should be selected in advance, and the guns hurried forward to it, not necessarily in a direct line, but diverging to right or left if the change of line gives a more protected approach. It will often be possible for a machine-gun officer to go forward, select the new position, and signal to the section to come up to it.

Of course the guns cannot be continually firing. There will be frequent pauses in their action, and at the outset and again after each advance in the first stages of the attack they may be for some time waiting for a favourable target on which to open fire. This time can be usefully employed in cleaning, oiling, and testing the guns, improving the cover, and in all cases, as soon as a position is occupied, the range-finder will be set to work taking ranges on all the points on which it is likely that fire will be opened. The fullest use should always be made of the map to visualize the reverse slopes of hills. It will be useful for the range-finder to accompany the officer who goes forward to select the machine-gun position.

In an attack that lasts for some hours, the aggregate time during which a section of machine guns is in action may not be many minutes. The important point is that these bursts of fire should be as effective as the skilful handling of the guns, the choice of the target, and the taking of the range can make them. As the attack presses forward and the range shortens, the chances for effective fire will become more frequent. The guns will probably do their principal work at medium and short ranges, and care should be taken to have a good supply of ammunition in hand for this later stage of the attack. If by any chance there is a deficiency in their ammunition supply, it is better to work the guns silently forward without coming into action in the earlier stage of the advance, and reserve all their power fire for the later and more critical stage. *The successful action of the guns depends primarily upon careful arrangement of the ammunition supply.* This is one of the reasons why there would be a distinct gain in having staff officers attached to every division and brigade, whose duty it would be to assist in the general direction of the machine-gun work, watch the movements of batteries and sections as they push into the fight, and see that ample reserves and ammunition are sent forward after them. With a good system of supply, the officers actually commanding the guns in the fight ought to have no anxiety about husbanding their fire and sparing cartridges at the moment when the situation calls for their utmost efforts.

It is laid down that at long ranges the fire of machine guns, and especially of a combined group, is effective against such broad and deep targets as is supplied by massed bodies of his troops. But in actual practice such targets will very seldom present themselves. The chances are that at the beginning of the advance the machine gunner may see nothing of the enemy. What he is most likely to see is merely the ground where, according to good report from the air service, the enemy is known to be in position. It is obvious, therefore, that whatever the theories about long-range fire may be, there will be little chance of its employment.* Nevertheless, if such a chance presents itself, it must not be lost. And this is why it is suggested that the guns detailed to accompany the attack should have already occupied a firing position as soon as long-range fire is opened by the infantry.

At this stage, however, the infantry fire will be more effective and the machine guns will be silent, unless the enemy shows a favourable target. With good leadership on the opposite side this will rarely happen, but men make mistakes, and if the enemy gives the machine guns a chance they should

* That is to say, by the machine guns accompanying the attack. Action against close reserves of the enemy will be left to the guns detailed for long-range covering fire.

not lose it. The more effectually their position is concealed, the greater is the chance that somewhere within range of their fire a careless enemy will give them their opportunity.

When fire is opened, as a rule, it will be direct. Chances may present themselves of opening a long-range fire upon the enemy's supports, but as a rule this will be better left to the machine guns detailed for covering fire and placed further back. The guns accompanying the attack will have to deal almost entirely with the enemy's front line. Their real work begins as the range shortens, targets become more visible and observation of fire is easier. With guns in action within a thousand yards of the enemy's position, the first trajectory of their fire almost excludes any kind of indirect firing.

In arranging an infantry attack it is usual to assign a certain frontage to each unit, and in the battalion to each of the companies put into the firing-line at the outset. The company works forward on this front, and has for its objective a similar frontage on the enemy's line. But the machine gunner finds himself acting under somewhat different conditions. His guns, whether they are a section or a company, occupy a comparatively small frontage, and he has to look for his targets not necessarily in his immediate front, but it may be to other points to the right or left front within reach of his guns. And the firing-line itself is no longer unbroken by intervals. At first the tradition of the old shoulder-to-shoulder methods led to there being considerable hesitation about leaving any intervals or gaps in the advancing line. *But now it is recognized that such intervals may be safely left in attack as well as defence, provided their front can be swept by fire from the right or left.* In case of such an interval being left in the line, or developing as the attack proceeds, the gap marks a useful place for the machine guns. They need not be actually in it, though, as a matter of fact, the unoccupied ground will often give them an excellent choice of covered fire position; but if they are not in the gap, their place will be near the flank of the troops to the right and left of it, and it will be part of their mission to watch this part of the front.

The best targets will be afforded by men in movement on the enemy's side. One can hope for very little effect against an extended firing-line under fairly good cover, but there will almost invariably be moments when reinforcements pushing up to the firing-line, or parties of men bringing up ammunition, will show themselves sufficiently to present an easy target. But it is above all when the enemy attempts a counter-attack that the machine guns will have their chance. The forward movement through the enemy's advanced line will bring into view numbers of the enemy who, however well trained they may be, will not be able to find cover on the direct line of their advance. In such a case the sudden opening of a heavy fire by the machine guns may bring the counter-attack to a standstill at a very early stage.

As the attack proceeds and the range decreases, the opportunities for the action of the machine guns will become more frequent. At long range there can seldom be any opportunity for the oblique or enfilade fire, which can be delivered with such effect from a machine-gun section, but as the attack breaks into the enemy's advanced line, the opportunity for enfilading trenches or lines of cover will frequently present itself.

There will also be fights for special points of vantage, and in these local encounters, where movement will often be restricted to well-defined lines and narrow fronts, the special powers of the machine gun will come into play. In the attack on buildings of most kinds it is true that the stream of bullets is ineffective against the cover afforded by solid brick or stone walls. *But it is particularly deadly against windows and other openings from which the enemy's fire is directed,* and it will usually penetrate roofs of all kinds, thus driving the defenders from the upper parts of the buildings attacked. In these fights for farmsteads and villages the gun comes into play in many ways. In the attack on barricades its fire can sometimes penetrate improvised cover, and in any case, if used with a short traversing movement, it can sweep the crest of the obstacle and beat down the enemy's fire, unless he is protected by very carefully constructed head-cover with good loopholes.

Again, the guns are of the utmost value in helping to secure each point that is won. This is an individual application of the general rule that when a position is stormed immediate steps are to be taken for securing against counter-attack, and the easiest way of organizing the defence against the counter-attack is to bring the machine guns up.

One may say, indeed, that the most important time for the machine gunner is immediately before and immediately after the assault, either upon some feature of the enemy's position or the position as a whole. In the outburst of close-range fire which is the final preparation for the charge, the guns must be brought into action and worked to the extreme limit of their power. It is the moment when it is important that their well-recognized effect in acting on the spirit of the attack and breaking the nerves of the defence should be used to the utmost. And their intervention has a special value from the fact that quite apart from its volume their fire will probably be much more accurate and effective than that of the infantry beside them. A close examination of the effects of fire in the battles of the Manchurian War has brought out the unexpected result that both with the Russian and the Japanese armies at extremely short ranges—150 yards and under—rifle fire was comparatively harmless. At such short distances one might expect that it would be deadly, and no doubt it would be extremely destructive if on both sides fire was first opened at this short range. But by the time the short range is reached the men on both sides have been subjected for hours to the strain of battle. They are tired and excited, and, unless in the case of very exceptional men, they no longer shoot steadily; they have a tendency to throw up the rifle on bringing it to the shoulder, and almost all the fire is too high and passes over the opposing line. *It is in this stage of the fire-fight that the nerveless machinery of the gun sends out the steady stream of bullets with no variation from the elevation fixed by the gear on its stand.*

This mechanical accuracy of fire also enables it to be kept up with safety to the troops that it is assisting much longer than anyone would risk keeping the fire of covering bodies of infantry in action. The fire of the guns, in fact, need only cease when the final rush with the bayonet begins. Then comes the moment when the officers in charge of the reserve machine guns should be ready to move forward in rear of the foremost wave of the attack. It is not merely that they will thus be ready to assist at once in organizing the defence of the ground as soon as it is won, but they may have opportunities of intervening suddenly and effectively in the actual fight for the position.

Their most important task will be the securing of the captured ground. Everyone recognizes that one of the most critical moments in a battle is that in which a position has just been stormed. The victorious troops are confused when disorganized by their own success, and even with the best-trained men the work of rallying and reorganizing takes an appreciable amount of time. This is why it is considered advisable, if possible, *to move on to the ground a reserve of fresh troops. Here the reserve machine guns can be particularly useful.* They will go forward with the attack and can be rapidly placed in position, and be at once ready to bring a heavy fire to bear upon a counter-attack, or if the enemy is thoroughly broken and retiring, they can open a pursuing fire. The guns that assisted in the attack should be reassembled by sections and fill up with ammunition, so as to be ready to be sent forward in turn as a second reserve. It has been suggested that in breaking into an enemy's battle line the attacking force should be well provided with machine guns on its flanks. This is a sound idea, for the counter-attack on the flank is even more likely than an attack from the front. *It is in this rapid taking up of a position amid the confusion following an assault that ready tactical judgment and a good eye for ground are all-important for the officers commanding the machine gun.*

Application of Fire

As to the targets selected for fire in battle, one cannot always expect to have the ideal target, combining depth with front. It will often happen that guns will have to be brought into action against what is theoretically a bad target—namely, a hostile firing-line in very open order. To obtain a good effect in such a case, the action of several guns will have to be combined, first to find the target and then to bring a considerable part of the skirmishing line simultaneously under fire by *placing the descending bullet sheaves of the guns side by side along the hostile front.* This will give a prospect of suddenly sweeping away a whole sector of the enemy's fighting line, and will produce a greater effect than a mere spraying of the line from right to left by traversing fire. It is an application of the principle that the greatest moral effect is produced by sudden and serious losses.

In Russian narratives of the Manchurian War one finds some instances of effective machine-gun action against a Japanese firing-line by the method known in the Russian Army as "zone fire with sweeping." This consists in firing a certain number of rounds at varying elevations, at the same time traversing the gun from left to right and back again, the object being to cover a rectangular space with a hail of bullets. In an article published in the *Revue d'Artillerie* in January, 1905, which gave long extracts from the diary of a Russian machine-gun officer, published in the *Russki Invalid*, there is an account of the dispersion of a Japanese attack in open order by this kind of fire. In the same narrative we have an account of the destruction of a Japanese battery on the move. In this case combined sights were used with four guns, the first firing at 1,200 paces and the others increasing the range each by 25 paces. The destruction of the battery cost 6,000 cartridges.

In discussing the use of machine guns on the battlefield one naturally thinks first of the attack, but in every battle they will also have to be used on the defensive on certain parts of the line, and even where an attack is in progress there will continually be periods when the guns have to be used temporarily and locally on the defensive. In fact, one of the chief uses of the guns will be to provide an ever-ready reserve of fire to hold the enemy, break up a local counter-attack, and secure the ground already won.

Gun Intervals

In both attack and defence there should not be, and there need not be, any crowding together of the guns. As a rule even the two or four guns of the section should not be close together, and the massing of sections side by side is generally a mistake. By keeping them apart, there is less chance of their becoming easy targets for hostile artillery fire, *and they can combine their own fire*, even from widely dispersed positions —in fact, dispersion is an advantage, for thus the guns can more easily cross their fire, and oblique fire is generally more effective than frontal fire.

Rearguard Action

In case of the temporary or complete failure of the attack the machine guns will have to do their utmost to cover the reorganization or retirement of the infantry they are supporting. Some of the guns will be used to create a screen of fire between the infantry and the enemy; others will use high-angle fire against the reserves which the enemy is bringing up for the counter-attack. Work of this kind is practically rear-guard action. In a prolonged retirement a rearguard should be well provided with machine guns. In such a case it will be well to place a number of sections or companies under one commander, who will supervise their general movement, and the best plan to adopt is to divide the machine-gun force into two portions, one of which will be in action, while the other is taking up a second position farther to the rear. The guns of the first detachment will retire through this second line. All retirements can be made gradually, a few of the guns being kept in action until the last moment.

CAVALRY

Cavalry are now so largely used in the dismounted combat that the action of machine guns attached to them will generally be the same as infantry. There have been in the present war few instances of cavalry charges, even against hostile cavalry. In the case of cavalry engaged against cavalry, the action of the supporting machine guns will be analogous to that of horse artillery. They will endeavour to take up a flank position, from which they can bring their fire to bear upon the enemy's cavalry while their own cavalry is advancing to the attack. If such an opportunity occurs the machine guns will always have an ideal target in the close-ordered lines of mounted men forming the enemy's force. The guns will have to provide for their own protection, and will often be able to take up a position where they are practically safe from mounted attacks.

OUTPOSTS

With regard to the employment of machine guns for outpost duty, all that need be noted is that they will often be useful where a picket is watching a road, bridge, defile, or other marked line of approach. In such a case arrangements can be made for night-firing. Where the outpost line is also intended to be the line of resistance it will be useful to have a larger number of guns with the outposts, and to select in advance the positions to which other guns are to be brought up.

Aircraft and Motor-cars

Without attempting to do more than give some general indications as to the use of the machine gun with aircraft and with motor-cars, something must be said of these recent developments.

The petrol-driven motor-car was still in a very elementary stage when it was suggested that in coming wars it would be armoured and armed with machine guns to make dashing raids along the roads on an enemy's front and flanks, or even far in his rear against his communications. It would be something like a land torpedo-boat. Sanguine advocates of the motor-car in war even made forecasts of armoured cars moving across country charging into the midst of hostile troops and destroying them with the all-round fire of machine guns and light quickfirers.

It was some time before the motor-car became sufficiently reliable for even much more modest claims on its behalf to be accepted by practical men. But before the Great War began it had been so improved that it had been adopted by most of the armies of Europe for a great variety of purposes, notably for the conveyance of Staff officers and messengers, and for replacing horses by mechanical transport. It has also been used on both sides for the rapid bringing up of infantry, either to reinforce a fighting line or to support the advanced cavalry. With these developments came the armoured fighting-car with the machine-gun armament.

In the British service the armoured car was first introduced by the Navy. It was the result of the occupation of Ostend in the last week of August, 1914, and the establishment of a base there by the Royal Naval Air Service. They found the Belgians in West Flanders were using improvised armoured cars on the roads, and some of these cars proved to be useful helpers in the work of our airmen. A report from Ostend to Mr. Winston Churchill, then First Lord of the Admiralty, led to the immediate organization of the Armoured Car Division of the Naval Air Service.

With Ostend, and subsequently Dunkirk, as its base, the Armoured Car Division did excellent work in reconnoitring in connecting with the airmen and the cavalry, and in driving in the enemy's patrols in the district. The sailors handled the cars like "land torpedo-boats," to use an expression of Lord Charles Beresford. Their first exploit was thus officially chronicled in an Admiralty Report, dated September 18, 1914:

> On the 16th instant Commander Sampson, with a small armoured motor-car force attached to the Naval Flying Corps, encountered a patrol of five Uhlans, near Doullens, killing four and wounding and capturing a fifth. The British force suffered no casualties.

After these first experiences armoured cars with machine guns were attached to the British forces, not only in France and Flanders, but also in the Gallipoli Peninsula and in Africa. They even took some part in General Botha's operations in German South-West Africa, moving on very different ground, and in one instance bearing the chief share in the repulse of a German attack on the British camp near Wallfisch Bay. In the Gallipoli Peninsula they were used with effect on almost roadless ground.

As a general rule the heavier type of armoured car, in which the gun is permanently mounted in a kind of turret, will operate on the roads. It is obvious that it can give invaluable support to patrols and detached parties, the men working the gun being themselves fairly safe from rifle and machine-gun fire. The drawback of the car is that it is rather a prominent target for hostile artillery. Here its power of rapid movement is its chief resource. The commander of an armoured car has to run the risk of being knocked out by a direct hit, and the further danger of being ambushed, with the road which is his line of retreat being obstructed behind him at the same time. The Germans have repeatedly tried to dispose of a patrolling car in this way, but it is not certain that they have ever succeeded.

There is another type of motor-machine-gun mounting evolved from the motor-bicycle and its side-car. The gun, tripod, and shield, and a first supply of ammunition, are conveyed in the side-car. In this case the motor brings the gun up to the scene of action, and it is then taken from the car and placed in position. It could be actually fired from the side-car, but for practical purposes the motor and car should be regarded rather as the means of bringing a reserve of guns at high speed to the place where they are wanted.

Motor-machine guns of both kinds have been successfully used to close a gap in a line or support troops who are rapidly being driven in. It is obvious that in covering a retreat along a good road they would have a very high value. We may take it that their normal work will be to support patrols and advanced parties, to operate against the enemy's patrols, and to form an efficient mobile reserve. The heavy motor-car has the advantage that, besides carrying the gun, it can convey a large supply of ammunition, and is thus an independent unit.

The experience of the present war shows that armed motor-cars can be used even on rough ground where there are no good roads. One of the armoured car squadrons of the Royal Naval Division did good work during the fighting in the Gallipoli Peninsula, and armed motor-cars were used in the desert fighting against the Senussi Arabs on the western frontiers of Egypt in the present year.

It has been proposed that machine guns mounted on

motor-cars, so as to fire at a high angle, should be used against hostile aircraft. The proposal seems attractive, because the gun could follow the enemy's airmen and often move as rapidly as they can. The drawback, however, is that against an aeroplane no serious damage can be done, except by a lucky hit, on a very few points. To kill or seriously wound the airmen, to damage the propeller or the engine, or cut some of the wire stays, might dispose of the aeroplane. But the mere riddling of the planes with bullets makes practically no difference to the flying power of the machine. Against dirigibles the mere firing of machine-gun bullets is still more ineffective. In a paper on aerial warfare read by Mr. Walter F. Reid at the Royal United Service Institution in February, 1911, the lecturer said:

> So far the effects of artillery fire against balloons have been disappointing. Captive balloons, which wait patiently until they are hit, have indeed been brought down, but even if hit they are not necessarily placed *hors de combat*. Colonel F. C. Trollope mentions that only one case occurred of a balloon being hit during the Boer War. A shrapnel shell fired at a range of about 600 yards burst in front of it and made sixty-eight holes. But the balloon took twenty minutes to come down, and was subsequently mended.*

This incident is quoted here to show how little result can be expected from machine-gun fire against the huge aircraft of to-day. If a captive balloon suffered so little damage from sixty-eight penetrations, it is evident that very little damage would be done to a Zeppelin by a large number of machine-gun hits, for these giant airships contain within their rigid case fifteen or sixteen independent gas chambers, and the riddling of one or two of these by machine-gun fire would lead only to the slow escape of gas from the damaged compartments, and thus diminish only to a relatively small extent the lifting power of the airship. Machine guns have been used, and are being used in the present war, against aircraft, and

* *J.R.U.S.I.*, June, 1911, p. 742.

no doubt it is useful in so far that it endangers the lives of their crews and forces the airship to keep at a high altitude, but it is clear that the real weapon against the airship is the gun mounted at a high angle and firing shells.

In the early days of aircraft—and these early days were only a few years ago—it was only a small group of enthusiasts who ventured to propose that an aeroplane could be armed and converted into a fighting as well as a scouting unit. As late as February, 1911, in a discussion on the subject at the Royal United Service Institution, more than one expert protested against the idea as impracticable. Before the war machine guns had been mounted experimentally on a few aeroplanes.* It is evident that the machine gun, which absorbs its recoil in actuating its mechanism, is the ideal weapon for this purpose. Rifles have been used from aeroplanes, but the machine gun is a far superior weapon, the rapidity of its fire increasing the chances of a really successful hit, and the machine gunner on an aeroplane has a better chance of damaging a hostile aircraft than if he were firing from the ground. He can hope to obtain a position from which he can bring his fire to bear on the vulnerable points in a direction from which more than one of them will be within his possible line of fire, and he can close to an easy range. The discussion of the best method of action—the tactics of the machine gun in aerial warfare—must be left to the airmen. Much interesting information on the subject will no doubt be available when the end of the war comes and they are able to speak freely about matters that are now necessarily kept secret. Meanwhile, in order to give some idea of what aerial fighting is like, we may give an extract from the work on "Aircraft in

* In June, 1912, the Lewis gun was well tested from a two-passenger Burgess-Wright aeroplane at the U.S. Army Aviation Station at College Park, Maryland. On November 25 and 26, 1913, Captain K. R. Davis fired a Lewis gun from an aeroplane at Bisley at targets made of sheets 2 feet by 30.

the Great War," by Claude Grahame-White and Harry Harper, published in March, 1915:

There were, at the outbreak of war, a certain number of French aeroplanes, with motors of 140 h.p., which had machine guns fitted to them. But these craft, owing to their power and speed and the weight they carried, were difficult to handle; none but an expert would dare to fly them, while the risk of damaging them in alighting was so great, owing to the pace at which they made contact with the ground, that few survived for long the rigours of active service. But the need was such that supreme efforts had to be made; and before long French pilots were given a biplane, steel-built throughout and with large sustaining planes, and fitted with a motor of 200 h.p.; and the machine proved so efficient that, in spite of the weight of machine gun, combatant and pilot, it would attain a speed of more than seventy miles an hour. This type of craft, of which as many as possible were sent on active service, has done admirable work. British gun-carrying craft, also, have been greatly improved, thanks to the experience of the war.

The value of a machine gun in an aerial combat, with its comparatively long range and the concentration and rapidity of its fire, was shown by a fight which took place between a French biplane, flown by M. Louis Paulhan, and a Taube monoplane, steered by one of the German pilots. With M. Paulhan was a passenger, whose duty it was to handle the machine gun, the airman contenting himself with the piloting of the machine. They were on a reconnoitring flight, passing high towards the German lines in the direction of Amiens. Below, here and there, floated a film of cloud. Suddenly, beneath them and to the rear, appeared from the clouds the German monoplane. Paulhan, quick to realize the advantage that his height gave him, swung his machine in a half-circle and dived like a hawk above his foe, bringing himself in one rush to within 500 feet of his enemy. But the German pilot also was a man of action. To continue on the course he had been holding was, he saw, merely to court destruction, seeing that he was in a position of tactical disadvantage. So, making a quick turn, and diving to increase the speed of his machine, he attempted to avoid the encounter and swung away upon his opponent's flank; and, had the weapon of the Frenchman been a rifle or revolver, the German would certainly have escaped. But as it was, opening fire promptly with his machine gun, the passenger in Paulhan's craft, having for the moment a broadside view of his enemy's machine, riddled it with bullets and sent it crashing to earth. A stray shot from the German, fired just before he was put out of action, pierced the petrol tank of Paulhan's machine; but the aviator, flying back towards his own lines, was able to land safely near a French battery.

In another instance, which revealed the effectiveness of machine-gun fire, a German biplane was passing above the lines of the Allies when a French craft rose unexpectedly to meet it. The French pilot steered his machine straight at his antagonist; while his passenger, opening fire with a machine gun, was able not only to puncture the German's petrol-tank, but to shoot dead his passenger. The German pilot began to plane earthward; but, suddenly owing to a leakage of petrol, the aeroplane burst into flames, and he was burnt to death before he could escape from the blazing craft.*

Machine guns are said to be mounted on some of the German aircraft on a platform on the top of the body of the ship, with a view to supplying fire effect against hostile aeroplanes. Such an armament will give a very limited protection against a daring adversary. Probably the true defence of an airship against aeroplane attack will consist in its being escorted by aeroplanes, as a battleship squadron on the sea has its escort of destroyers.

The most recent and the most authoritative work on aerial warfare is a book published in the present year under the title "Aircraft in Warfare," by Mr. F. W. Lanchester, a member of the Government Advisory Committee for Aeronautics, with a preface by Major-General Sir David Henderson, the Director-General of Military Aeronautics. Mr. Lanchester considers the Lewis gun the best existing armament for the fighting aeroplane. He suggests, however, that it may be worth while to devise a machine gun throwing a heavier round-nosed bullet, which at short ranges would do more damage to a hostile aeroplane than the present small-bore pointed service bullet, and that possibly two or three barrels might be combined in one gun to increase the volume of fire during the brief time when the gun is in action in a favourable position. Lewis guns in the Air Service are fitted with larger

* "Aircraft in the Great War," p. 329.

sized magazines containing forty-seven cartridges. This is done to save the time that would be spent in changing the empty for a full magazine. In aeroplane work it is fairly easy to keep the guns cool; the rush of air caused by the motion of the aeroplane gives a superabundant air-cooling action. Even when Maxim guns are used the water-jacket is generally removed.

Guns are now often mounted in aeroplanes in a fixed position without any pivoting. Aim is then taken by directing the aeroplane itself towards the target. The gun is mounted in front of the machine, and fires through the revolving propeller.

CHAPTER X

TRAINING

MACHINE guns can only develop their true powers in the hands of thoroughly trained gunners. The officers who command them must not only be themselves masters of their weapon, but must also have sound tactical knowledge, an eye for ground, the faculty of quickly appreciating a military situation and the habit of forming prompt decisions.

In the hands of half-trained officers and men the machine gun becomes a mere waster of ammunition on a portentous scale.

Sound training is therefore all-important, and it must have a very wide scope.

There are two marked stages in the process of training. Some people seem to think that the work is done when the first stage has been completed; but essential as this is, the second is of even greater importance.

In the first stage of training officers and men have to learn to know the gun thoroughly. They must acquire the utmost familiarity with its mechanism and working—playing with it, pulling it to pieces, assembling it again. They must understand how every part of the mechanism does its work. They must acquire the power of substituting a spare part blindfold or in the dark by mere touch—for such work has often to be done in absolute darkness on service in the trenches. At the same time they must get an intimate knowledge of their gun's powers: first, theoretically, then by varied practices on the range and over broken ground. They must know at once when anything is going wrong, without having to look at the position of handles or hunt up formulæ in a book. The gun is a machine, and whoever runs machinery has to know when anything is wrong almost instinctively, by sound and touch, and be able to set it right at once.

The drill of the gun must be known thoroughly. Whoever commands must give the right order by mere force of habit, and the men should execute it with the swift certainty that is only acquired by continual practice.

Officers and sergeants should know all about ammunition, spare parts, mountings, stores, etc., and be able to do everything that the men do, and, if possible, do it more swiftly and efficiently. They must be able to teach by showing how the thing is done, watching the men doing it and putting any mistake right at once.

But all this is the mere ABC of the matter—work to be done in the barrack-room and on the drill ground and the range. It is the necessary foundation, the first stage of training. But the second stage, the more important and difficult, begins as this ends.

Those who direct the training of men in the first stage should have already advanced far into the second. One cannot say that they should have completed it, for this higher training goes on all through a man's working life. If he is keen and efficient, he is always learning. The teacher in the first or elementary stage should fully realize the bearing of all that is done upon the actual work of war.

The training should not be hurried. In the first stage a great amount of hitherto unfamiliar detail has to be taught, not theoretically, but practically, and good practical training means the gradual formation of a habit, so that the right thing is done without thinking about it. This means time. Hurried training is necessarily defective, though under present conditions it sometimes has to be used as a makeshift. It is unfortunate that this should be so, for we are only too likely to get partially trained gunners whose training has to be completed by the experience of actual service. This is rather a wasteful method, for too often it will mean throwing away a lot of ammunition, losing a number of men in action unnecessarily, and sometimes failing to produce any useful effect.

The second stage of training is for those who are to command. Its object is the acquisition of a sound practical knowledge of the handling of the machine gun in battle.

In this matter correct theoretical knowledge is all-important. There is sometimes a tendency to despise theory and to trust entirely to the rough-and-ready knowledge that is acquired by practice. But, as the great von Moltke once put it, "Though there is a long step from theoretical knowledge to actual practice, there is a vast abyss between the want of knowledge and useful practical work." Knowledge derived from personal experience is most valuable. But it is much too costly if it has to be acquired at the risk of a series of dangerous mistakes, which might have been avoided if one were guided by the experience of other men; and sound theory is nothing but the summarized results of practical experience.

Some twenty years ago the late General Sir Frederic Maurice, in a paper on "The Study of Military History," referred to the prejudice that exists in many men's minds against the theoretical study of war, and especially of military history. He pointed out the importance of preparing for one's own practical work by taking advantage of what could be learned from those who had already done work of the same kind, and he went on to say:

> There is something almost comic in the attitude of mind in regard to this subject of half the English soldiers and more than half the civilians one meets. If a distinguished soldier, like Sir Redvers Buller, goes down to Aldershot and describes certain war experiences of his own, they are delighted, interested, and absorbed by what he tells them. They declare at once that these are "worth all the theory in the world." Now, what I want to ask such men is, Would these experiences of Sir Redvers Buller have lost their

value, if they had been recorded in a printed book? Because if not, this horrible thing which you call theory is simply the same thing which so much interests you in the other shape.

He adds that however useful one man's experiences may be, no man's personal experiences can be so valuable as the compared and collated experiences of many men. And here again we have theory in its best form.

The reader will have remarked that, in elaborating our theory of what the tactics of the machine gun should be, we have freely quoted from the writings of soldiers of many nations, after having previously told how machine guns had actually been used in battle, and traced the evolution of the gun itself. We have here an example of what the study of any practical subject should be. It is important to realize how things have come to be what they now are. One has a more intimate knowledge of any subject if one has traced it from its origins, and it is important to have the views, not of one book or one writer, but of various minds, so as to see the thing from many points of view, and, above all, give due weight to the teachings of varied experience under different conditions. This is the meaning of our description of sound theory as a reasoned summary of practical experience.

For most men, and especially for young men, their own experiences will form the least part of their sources of knowledge. Theoretical knowledge is chiefly to be acquired by reading, thinking, and talking things over with others. And of these three methods, the first two are the most important.

Our "Training and Manœuvre Regulations, 1913" (Section 8), give an excellent summary of what the training of the officer should be—a training which largely depends upon himself. The following passage is well worth noting:

An officer's first duty is to fit himself to become a capable leader in war. But it is of an almost equal importance that he should be an efficient instructor in peace; for it is on the training given by him to his subordinates in peace that the war efficiency of his unit depends. He must have sound military knowledge, which can only be gained in peace by a study of how military situations have been dealt with in the past, and by constant practice in thinking out and deciding how best to meet definite military problems with which he is likely to be confronted in war. He needs imagination in order to have the power of anticipating probable future requirements and developments, and organizing ability in order to make the necessary provision to meet them. He must have a good eye for the use of ground, a capacity for grasping a situation in all its bearings and for coming at once to a correct decision, which he must be able to translate forthwith into clear and concise orders.

It will be noticed that the official manual here lays stress upon the need of constant *thinking* as well as *reading*. There is always a danger that a young officer, if he reads at all, will be content to run through the pages of a book, perhaps here and there committing to memory or making notes of a few lines which he thinks may be useful, but obtaining only vague impressions, and failing entirely to make what he reads a permanent part of his mental equipment. Mere reading is not study. Lord Wolseley, when he commanded in Ireland, put the matter very well in addressing an audience of officers on the study of military history:

> A certain amount of reading (he said) and a certain amount of study is absolutely necessary for any man who ever wishes to command troops in the field; and (he added) so far as I know of the study of war, the great thing is to read a little and think a great deal, and think of it over and over again. . . . I do not believe that this advice can be bettered. A few campaigns thoroughly studied will do more to strengthen the intellect, to develop a capacity for hard thinking, and to teach the art of leading troops, than fifty campaigns that have been merely skimmed.

What one has to aim at is to know things thoroughly—to know them so well that there is no effort of memory in calling up the knowledge, but it is at one's command almost unconsciously. It has to become part of one's habit of thought; and this cannot be done by mere reading.

As a guide to thinking things out, it is well to take up

this or that point, ask oneself what one already knows about it, decide in one's own mind where one's knowledge is doubtful or incomplete, and then turn to the books in order to verify and supplement what one already knows. This gives a definite object to one's reading. Some students have a habit of accumulating extracts from the books they are studying. This is useful enough, because the very act of collecting and making the extracts rivets one's attention upon the point in question. But it is better to supplement this practice by summarizing the matter in one's own words, noting perhaps here and there a telling phrase of one's authority, not merely copying (which is repeating the thoughts of someone else), but, besides this, translating the matter one deals with into one's own mode of thought, trying to make it part of oneself.

The two great British leaders in concise thought on fire tactics—Lieut.-Colonel C. B. Mayne, R.E., and Lieut.-Colonel N. R. McMahon, D.S.O.—are now dead, and the standard work by the former is out of print. (For information about these officers and their works see the bibliography in the Appendix.) Company commanders are recommended to read and reread Commandant J. Colin, formerly a professor of the French École de Guerre (or Staff College), whose work, " The Transformations of War," is now translated into English. This is a most useful work for close study, and helpful in a high degree for the working out of major fire tactics.

Private reading and thinking are absolutely necessary for all officers, and this must be kept up during the whole of their service. To put it in another way, officers must always keep carrying on with their reading. Where there is a will there is a way, and an officer can keep up his reading while on active service in many of the theatres of war. In the present work there is gathered together and worked up material which will enable several courses of reading to be carried out, while the great weight of a large number of books will be saved.

It must not be forgotten that the battle tactics of machine

guns are only a part of fire tactics, whether there are two or a hundred guns firing. One of the advantages of a Machine-Gun Corps should be a consolidation and recording of all thoughts on the subject, thus providing a broader foundation upon which to carry on the training of the officers and men in all its stages. While there is a very great deal of sound thought on the many parts of the art of war in the official manuals and regulations, yet it is absolutely necessary that officers have guidance in how to start and carry on the reading of these books. Reading circles constitute one good method of starting new officers or N.C.O.'s on the soundest way of reading the different training manuals and the field service regulations. The members of the circle sit round a table, each with his own copy of the book to be used, while the instructor details the lines to be read aloud by each member in turn. He stops the reading at any point he thinks should be emphasized, which he does in a few words, and he may ask the reader for his own ideas on the point.

There is a great omission in the official textbooks, and that is the necessity for "mystifying the enemy." The word "mystify" is not to be found in the official textbooks. The nearest approach is "surprise," which is not nearly so comprehensive nor emphatic. Even the entries under "surprise" appear to limit its meaning to the movements of *troops*, and not to touch the movements of *projectiles*. As in the case of troops, so with the movements of projectiles, endless combinations can be worked with time and direction. There is a wider suggestiveness in "mystifying the enemy" than in "surprising the enemy," and it is possible that much of the objection to suppleness existing in British tactics in divisions and brigades in the present war can be traced to this omission. This suppleness and adaptability in thought should be handed down through commanders of army corps, divisions, and brigades, to those of battalions. Its lack can be traced to the inadequate grants for military training in the past, to the small size of the staff college, and to the absence of encouragement for all officers to take part in discussions. Young officers should be encouraged to think about the art of war from their first year of service. They should begin this by building up a store of thoughts from the digested records of examples from the past. The habit of discussion after a lecture or a tactical exercise should be carefully nursed, year by year, during the whole service of each officer. In order to succeed in this, commanders of units must never grow old in thought, and should take a very wide view of training. The habits of suppleness, adaptability, and mystification, must be formed while young, and strengthened and maintained all through an officer's service. He must read and reread, think and think again; but all this is of no use if he does not speak or write his thoughts. If he does not read military history, he will perhaps be puzzling over questions that others have solved long ago. By reading military history he can start his line of thought where the finest soldiers of the past left off; but to do this he must have good guidance. An officer instructor, whoever he is guiding in reading, must first get in sympathetic touch with his pupil. Probably an hour or two at the start by the instructor will guide the pupil into the correct path for most of his service, but this depends on the instructor. All this only deals with individual training of the officer. There are very suggestive thoughts on this subject to be found in "Training and Manœuvre Regulations, 1913," chap. ii., which all officers should carefully read and reread.

While there are endless ways of "mystifying," yet guidance can be given in this subject, and examples are here given from the present war. One way of "mystifying" is to execute tactical events out of their official sequence. Another is to carry out certain tactics at the same hour for many consecutive nights, and spring a change on the last night, using machine-gun curtain fire to prevent the enemy's front line from being reinforced. "Fire tactics and the use of ground" cover a wide region of study. One way to work up the official textbooks is to read through the index of each and mark each reference which contains a thought on "mystifying" the enemy or "fire tactics and the use of ground."

The key to an efficient machine-gun company is thorough individual training of the N.C.O.'s and men. A section commander must know each of his men thoroughly, and what things he can do best. The first thing to do with a new man is to get his brain into such a condition that he responds to a simple technical question by a direct reply.

What has been stated about the individual training of officers applies with equal force to that of N.C.O.'s. In guiding the reading of the latter care should be taken to illustrate the dry thought in the official textbooks with practical examples from military history, and thus introduce the intense reality of the human element. The officer who has prepared himself by studying military history can guide his N.C.O.'s through the textbooks and help them to strengthen their thinking power. Each N.C.O. will probably require individual coaching as a foundation upon which to start his study.

In field training one of the first things an instructor has to learn is how to turn the mistakes of his pupils to good account in further training. As in all other training, this must be taken step by step. All barrack-square training must be clearly connected by steps to the most elementary field training; and in apportioning the latter the capabilities of the pupil must not be overstrained with too many steps at once. A constant fault in machine-gun work is overcrowding the guns, and this arises from a fault of the instructor.

From the first lesson the individual machine gunner must be made to work at adequate intervals and distances from each other member of the section. In elementary field training the instructor should be at least twenty yards from the gun, scout, range-taker, limbered waggon, or pack-horse, as the case may be. In more advanced field training the work can be supervised and directed from a distance. When correcting mistakes, try to be natural with the man, and get him to understand *why* he is wrong in his action.

The regulation standard test should only be considered as the foundation upon which to build a highly efficient machine-gun company, and not as the ultimate end of the training. It cannot be too much impressed on all officers that a very thorough individual training is the only true foundation for collective training in companies and sections. This can only be done by an officer taking trouble with each man individually from time to time. Provided with such a foundation as this the section officer can work his men effectively and without crowding.

One takes it for granted—though, perhaps, it is in some cases a hazardous assumption—that the machine gunner has a good knowledge of tactical theory generally. His special business is to develop his tactical thinking in the direction of finding out how machine-gun work fits in with tactical work generally. This should be his hobby. Of course, every other branch of knowledge in connection with machine-gun work should be a constant interest to him, so that he is on the look-out to learn anything and everything that can be of use to him, whether by reading, seeing, or talking with others. All this knowledge will class itself under one of four heads:

Material.	Tactics.
Organization.	Training.

Of these, the tactical head has the widest scope. It means the real work of the machine gunner.

He will, of course, be familiar with our own regulations. These are, for the time being, the accepted doctrine which is the general guide for himself and for those with whom he is to co-operate. But one must not think that they represent the last word on the subject. The whole thing is still largely in the experimental stage. Indeed, its special interest arises from the fact that the machine gun is as yet a new element in war, and all who handle it intelligently have to be prepared to work out new problems and face new situations.

This is one of the reasons why study and thinking are so important. Mere "rule of thumb" is no use in this kind of work. Resource and originality are continually needed; and if a man is not well equipped with practical and theoretical knowledge, he cannot safely venture to be original.

When one talks of study and remembers how much thinking is a part of it, it follows that the very least part of study will be mere poring over books. If a man has once acquired a thorough interest in his hobby and the habit of thinking about it, he can be learning something in every ride and walk. Tactics are essentially the handling of men and their weapons under varying conditions of ground; and it is easy to acquire the habit of looking at any stretch of ground and at once beginning to think out a self-suggested problem of how one would use machine guns in attack or defence in country of this kind.

Knowledge of ground is all-important. Without it one can neither choose good positions for the gun, pick out effective targets, obtain a good field of fire, secure cover while in position, nor select lines of advance and retirement. Some suggestions on the study of ground will be found in a special appendix to this chapter devoted to the subject.* Here we need only note that the subject includes map-reading and the acquisition of a good eye for the actual ground.

The power of map-reading should be developed by prac-

* This essay of the subject is the work of Major Longstaff, and is largely based on experience of the study of ground in new countries such as British Columbia.

tice, by the comparison of maps and ground, so that one is able to make a rough mental picture of the ground it represents, as one looks at the map. It is a useful practice when one is going over the ground, map in hand, to ask oneself what the country will look like as one reaches a new point of view, as one comes to a turn of a road in close country, or reaches the crest of a hill. One is then able to correct one's impression by comparing it with what one sees as a new stretch of country comes in sight at the fresh view-point. One should be able rapidly to pick out the form of the ground from the contour lines, making a rough estimate of what places are visible or concealed from a selected point on the map, and verifying this estimate by quickly making sections of the ground with pencil and tracing paper. Then the useful exercise may follow of taking the known trajectories of the machine gun, and finding out from what positions indirect fire can be used, so that the trajectory will clear the intervening obstacles.

When working over actual ground, it is important to acquire the habit of always keeping the north point in mind, whether by day or night, and being familiar with and constantly practising the ways of finding the north, not merely with the compass, but without it by the stars, sun, hands of a watch, etc.

Further necessary ground practices are the estimation of distances by the eye, without any instrument, besides the use of range-finders and the picking out of ranges by the map; combined with this there will be rough sketching of the ground and the rapid making of range-cards.

In most of our "little wars" on the frontiers of the Empire, we have had to work with maps on a comparatively small scale, supplemented by sketches made on the spot; but in the present war in France and Flanders elaborately detailed maps on a large scale are available. These maps are issued to our troops, divided by red lines into large and small squares, distinguished by letters and numbers, the sides of the small squares being a thousand yards in length, and these small squares being again subdivided into quarters, of which the sides are five hundred yards long. There is a further system of indicating a precise spot in these smaller squares by subdividing their sides into tenths. The system was originally adopted for the purpose of aerial reconnaissance and the direction of artillery fire, but it lends itself admirably to machine-gun work at the longer ranges. Maps of certain districts in England have been prepared with the same system of subdivision, and the machine gunner should make himself familiar with this method of indicating positions and targets on a large-scale map.

A further important subject of study and training is the question of cover—knowledge of the best way to make use of existing cover and of the most useful ways for improving it. Our experience in the present war, where the troops on the opposing fronts have been in close contact for months at a time on the same ground, under conditions that make good use of cover essential to avoid destruction or heavy loss, and where at the same time precautions have to be constantly taken against aerial reconnaissance, has led to a wonderful development of the whole art of securing cover and concealment.

In this connection it is well for the machine gunner to remember that with a little ingenuity he can often, not merely hide from the enemy the place where his guns are, but also mislead him into thinking that he has found them in a place where they are not. In the South African War one of our machine-gun officers repeatedly scored over the enemy by placing the old-fashioned carriage of his machine gun in a purposely badly concealed position, where the Boers were sure to spot it, whilst he opened fire with the gun from a tripod in a well-concealed position a couple of hundred yards away.

Again and again he had the satisfaction of bringing his gun effectively into action without drawing fire upon it, while the enemy were wasting their ammunition on the gun-carriage at a safe distance to right or left of him. This is a very simple example of what can be done in this way. It is a part of the too much neglected subject of "mystifying the enemy."

Machine guns will never act alone, and the study of combined tactical work is therefore essential. The machine gunner will always be helping other troops, and he must therefore have a sound knowledge of the tactics and methods of infantry, and, to some extent, of cavalry and artillery. He will have frequently to decide how he is to act without waiting for orders. He cannot do this, but will only blunder badly and do harm, unless he understands what is going on around him and where his intervention will come in most usefully.

A still higher tactical knowledge is required of senior officers, who have to control not one, but it may be many groups of guns. The divisional or corps commander will probably rely on the machine-gun commander to work out for him all the machine-gun arrangements and orders. This is the natural outcome of the specialization of command.

In connection with this training for combined work on active service there must be much more than parade-ground drill and firing at targets over the ordinary level ranges. Practical training should include field-firing over varied ground, first with the guns only, then with the guns working in combination with the infantry during their field-firing. On the average type of rifle-range, with its limited space and nearly level ground, field-firing is not of very much value.

Then there must be manœuvres with other troops with blank cartridge, beginning with manœuvres of a single section of machine guns with a company doing its field-training and ending with manœuvres of whole companies of guns with large bodies of troops. There are various difficulties in connection with the use of machine guns in a mere manœuvre battle. With the ordinary blank ammunition there is not recoil enough to work a gun of the Maxim type, and it has to be used as if it were a single loader. It is therefore not easy to show that it is in action. It is extremely difficult to make a reasonable estimate of the probable fire effect, or to induce an opponent to take account of it. Signalling to show the gun is in action has been used, but is not satisfactory. An essential point of machine-gun work is to keep the gun concealed, and it is rather absurd to put up or wave about a flag to show where it is, and that it is in action. The use of a rattle is also a clumsy and rather useless proceeding. Cartridges, specially fitted with soft wood, or papier-mâché bullets that break up soon after leaving the muzzle of the gun, have been used to work the recoil, but there is a certain amount of danger connected with them when used for blank firing. Probably the best plan would be to have a number of umpires detailed to watch the machine-gun work and to whom the machine-gun orders issued on both sides would be immediately communicated.

There is a general likeness in the machine-gun syllabus of instruction of most armies. These schemes are derived from each other, and differ only in minor matters of detail. The Canadian syllabus is here given, not as an ideal scheme, but merely as a kind of framework for some practical suggestions on various details of training. It is taken for this purpose merely because one of the authors* of this work has had experience of many kinds and systems of training. His notes added under each head have at least the merit that they are the result of actual practice and observation.

* Major Longstaff.

1. *Instruction in the Lewis machine gun.*
2. *Instruction in the Colt machine gun.*
3. *Instruction in the Maxim automatic machine gun.*
4. *Instruction in the Vickers automatic machine gun.*

A general description of the gun dealing with the *external* features. This also includes names of those parts of the gun which can be easily seen and handled without stripping.

Note that the whole subject of the mechanism of each type of gun is to be further dealt with under Sections 8, 9, 10, and 11.

In dealing with external features, begin with the muzzle and work towards the breech.

5. *Packing of limbered waggons and how to manœuvre with them.*

Speed and smoothness in unpacking and packing is especially important, and too much trouble cannot be taken in working out the detail for each of the waggon Numbers in the above drill. Each Number should know the articles he is to move and the sequence in which he is to handle them.

The manœuvring of the limbered waggons can be divided into—(1) Driving drill between white stones and posts; (2) Driving across country; (3) Concealed driving, or the approach to gun position through suitable ground features. All this training should be only carried out with *full loads* in both limbers; and for this any material will do, but earth or sand will damage the limbers least.

The use of the brake requires much judgment and practice. This Number should always apply the brake and "stand to" when the driver dismounts, so that the horses cannot run away with the guns.

6. *Fitting of equipment, guns, etc., on pack-horses.*

Pack transport is both old and widespread in the world. The two great essentials are *fit* and *balance*. By *fit* is meant the correct adjustment of the saddle frame, so that the pad on each side of backbone sits on top of the ribs and not on the sides. It follows that once fitted to a horse a pack-saddle and equipment should be kept solely for use on it, and not changed haphazard from horse to horse. Keep a note of number of horse and of saddle. The *fit* of a pack-saddle is more important than that of a riding saddle, as the cargo is a *dead weight*, and the horse cannot be so often relieved as in the latter, as a rider can dismount at any time. The good fitting of the saddle must be watched hour by hour, and at first it will be two or three days' work before the pads settle down on a new horse, and the stuffing may require altering. By *balance* is meant that the cargo on each side is equal in weight. If they are not, they must be made so in some way. In British Government Marks the cargo is secured by means of a surcingle, wanty-strap, or rope passing under the animal's belly.

Captain Morgan's Don'ts* will be useful to officers: "Don't, by rough lashing, move the load when once it is properly balanced. Don't tug or jerk, but pull steadily in lashing. Don't lose your temper because the panels keep 'coming down' during a first day's trek. They always do, and additional padding in front must be thrust through the stuffing flap in the panels. Ropes stretch, top bars come down, and cinchas (girths) get loose at first; but after the second day there is no further trouble. Don't fail to look well to the front top bars until the panels have settled; as a neglected top bar touching the withers causes a sore which needs a lot of curing.

"*Precautions Against Sores.*—See that front girth is well back from the fore-leg. See that the front of saddle is well clear of the withers when load has been on for an hour. If a sore appears, take out stuffing of panel over that spot, and quilt round the resulting hollow. To locate position of the sore, cover it with black grease, put on saddle which will thus become correctly marked. Always call a halt after first hour of a trek, to allow animals to stale and to look round the gear. Keep centre of gravity as low as possible. Don't girth up too tight. Don't tighten up harness on flat ground, as a well-balanced cargo on level country needs no cinchas (girth) at all."

The pack transport of guns, equipment, and ammunition is so important in machine-gun companies that every officer should be both efficient in its use and constantly practising the men of his section or company. The only satisfactory method of training horses and drivers is to take them out daily with full-weight loads. If the guns, etc., cannot be spared from standing drill, the weights must be made up, properly secured on each horse, and the horse taken an increased distance for five miles consecutively, beginning with four and ending with twelve miles, an increase of two per day. At least one officer should attend all five days from each section.

---

* "The Frontiersman's Pocket-Book," published by John Murray, 1911, p. 161.

7. *Construction of trenches and gun emplacements.*

Elementary training should consist in executing tasks by one section in daylight, when accuracy in carrying out the design should be insisted on, together with the working by shifts. *Advanced training.*—When a section can dig itself in effectively on the word of command in daylight, it should practise doing the same thing at night with no moon, working in shifts, to a time limit, and in silence. The siting should be done by an officer, and, if possible, in daylight.

8. *Stripping and assembling of machine guns (Lewis, Colt, Maxim, Vickers).*

This should first be done slowly and in daylight. Afterwards it should be done at night or blindfolded, and with a time limit. The complete stripping while blindfolded need only be done with the Lewis and Colt. All this blindfolded work is preparation for night firing on range. (Night firing must be carried out at night, and not by pretending it is night in the daytime.) All this should be part of the standard test, which after all is only the *foundation* of recruit training. A recruit machine gunner should be able to visualize in his brain the complete working of a gun firing. A recruit, officer or man, after picking up any small part of a stripped gun, should be able to give its name, place, and function, at once.

9. *Replacing broken parts of mechanism of guns.*

It is evident that if a machine gun is cold when fire is opened, it will take many seconds before the breech mechanism gets the same temperature as the barrel. The parts which most frequently become broken do so because they have not warmed up as fast as those immediately surrounding them. The striker is a frequent part to break in many systems; and this could be prevented if it was always kept at body heat and put in the gun just before opening fire. In cold weather a gun should be kept in a blanket coat.

The machine gunner will learn by experience that small parts of the mechanism, of which there are *no spares provided*, will get broken, in spite of every care. Therefore, each gunner should learn by heart the parts of the mechanism which are not provided for in the Spare Part Box. Care should be taken not to mix the locks of guns; each lock should only be used in the gun to which it is fitted and adjusted. The Maxim and the Vickers are the only systems in which an inclusive spare lock can be instantly replaced if damaged,

10. *The principles of construction and working of the mechanism of automatic machine guns.*

This covers a wider ground than the mere details of construction to be found in the textbook for each gun. The instructor should try to bring out the general principles involved; show how they are applied in the various types of guns; what is common to them all; where they differ and with what object; the difficulties to be met with in devising a practical gun, and the methods adopted in each type of gun to obviate these.

11. *Recognition and remedy of any defect of a gun during firing.*

In 1909, when the Author put his machine-gun section through their class-firing, he told them in the case of jams not to trouble about the position of the crank handle, but to *concentrate* on reopening fire with as few seconds' delay as possible. The section had been training for several years and knew the mechanism thoroughly. He made his own thick rims, bent cases, etc., and inserted three or four in the half belt for each man. Only Nos. 1 and 2 were at the gun, the rest of the section were twenty-five yards in rear. He was about six yards away half right with a stop-watch to time length of each stoppage of fire. The two men at the gun were absolutely alone and on their own resources. But each man made good, and the fire stoppages were very short. The men could visualize in their brain the working of the mechanism, and at once spot where the fault was in the cycle of movement. But the above will only be successful if each man has many hours in which to handle and rehandle the mechanism of whatever gun he is using.

12. *Use of the belt loading machine (Maxim, Vickers, Colt, Lewis, and Hotchkiss).*

This really includes the evacuation of the empty belts (Maxim, Vickers and Colt), drums (Lewis), brass clips (Hotchkiss), their transportation to the refilling point, their correct refilling, and their transport to the gun again. In regard to all canvas belts it is absolutely necessary that they be filled by the machine rather than by hand, so as to avoid faults in feed. The full detail should be worked out and written down for each system of feed, and practised with the regular Numbers and changed rounds, both by day and by night, against time, and on all kinds of ground. The position of the belt-loading machine must vary according to circumstances; it may be in a dug-out near the gun and far from limbered waggon, or it may be on the limbered waggon which may be near the gun. But the men must be drilled in both conditions.

13. *Range finding and adjustment of instruments (Marindin, Barr and Stroud).*

First and foremost this is a specialist's job—namely, that of a ground scout, who should also be able to sketch. Accuracy is the first duty in range finding, and after thinking the whole subject over one is surprised to find it is really a part of surveying, but the surveyor has to keep himself and his instrument hidden, and cannot erect an artificial beacon on which to observe. The ranges are really the "rays" of the topographical surveyor. The ground scout must know the exact error of his particular instrument under all conditions, and how, by means of three or more readings, to obtain the exact range.

There is just as much a necessity for ground scouts to be drilled and redrilled in the detail of their work, as for that of any other machine gunner. A fully trained ground scout should be able to do accurate work at lightning speed under all conditions of physical strength.

14. *Judging distance without instruments.*

Every machine gunner should know the keys to this by heart, and they are to be found in the "Musketry Regulations." There is really only one way to become efficient, and this is by continual practice. First become proficient, and then keep it up all one's service. In each distinct topographical and climatic area a trained machine gunner will have to revise and adjust his keys and standards of eye measurement.

Another point is that there are three factors in getting the bullets to hit the point desired: (1) There is the judging of the distance and its error; (2) there is setting the back sight to the estimated range elevation; (3) there is the error between the actual trajectory of the day and the tangent elevation. This also applies to No. 13.

15. *Visual training and increasing the vocabulary.*

Many men have a limited vocabulary for use in describing objects out of doors. This should be gradually increased, both simply and naturally day by day.

As regards the art of "seeing things," the eye of a game hunter is the most useful foundation upon which to build. He does not look for a sharp outlined form, but for a patch of a certain colour blended slightly with an uncertain background. This habit should first be established proficiently, and then kept up by a few minutes' practice daily.

16. *Concealment of the machine gun, and selection of sites for the same.*

This is chiefly a question of the "trained judgment" of the officer. It can be divided into—(1) Reading of ground; (2) use of long-range trajectory, etc.; (3) the co-operation required by the particular fire orders. His judgment can really only be trained by frequent exercises in solving the above problem on several kinds of ground. It is most important not to consider the siting of less than four guns. It is a common mistake to think of siting only one gun, as by doing so the asset of co-operation is gratuitously thrown away. The normal machine-gun fire unit should be four guns, and they should be so sited as to be controlled by their section commander, while of course concealment and disguise must be kept in mind.

The above heading should be put under one of a much wider nature but containing fewer words—"THE USE OF GROUND." This faculty will take many years to acquire and build up; in fact, one has never finished building up the faculty of using the ground to the best advantage to kill as many of the enemy with the smallest possible loss to one's own men. To become a master in the use of ground an officer must be daily thinking and working at it with his eyes; and this he must carry on for years.

17. *Qualifying in the official standard test for first-class machine gunners.*

It cannot be impressed too much on all ranks that the standard test is only a part of the *foundation* in the elementary training of machine gunners of all ranks. The whole of the drill comprised in the above must become second nature to each gunner, so that he can carry it out under all conditions of physical fatigue and mental excitement. It could with advantage be carried out twice: firstly, after a good night's rest, breakfast, and an easy march to the range; secondly, after being out all night, a rough breakfast, and a long march to the firing-ground in full equipment and heavy pack.

18. *Practice tactical schemes with ball ammunition.*

This heading covers a wide field of thought and action. Not less than four guns should take part in this training, and, besides having targets to "lay on," there should be targets laid flat on the ground, so as to show where the 75 per cent. of the bullet sheaf falls. The aiming targets should be in view of the layers over the sights, and the several kinds of fire should be practised: concentrated, dispersed, parallel frontal, combined sights concentrated, and parallel oblique. The section drill should be most accurately carried out, each man to his job, and no crowding round the guns or anywhere else. Due consideration to the use of ground should be most thoroughly given.

19. *Practice night firing with ball ammunition.*

The absolute necessity for this being carried out at night can only be adequately understood by those who have tried to do it on active service, after only having had night firing by pretence in the daytime. In the dark the simplest movement becomes difficult, and direct laying with accuracy is almost impossible. Of course, the Lewis can be used up to 400 yards, resting on a trench parapet, to stop enemy troops repairing wire, etc., as it is more of an automatic rifle and can be pointed as such in the dark. The most accurate laying for night firing is indirect, where the ground between gun and target is high enough to mask an electric torch.

20. *Practice indirect ball-firing with graticules.*

While the system of graticules is very rapid under certain conditions, yet the Author does not think the conditions are of sufficiently wide application to recommend its use. This method becomes inaccurate when the eye of the observer using the graticule card is more than about six feet above the gun. A different graticule card will be required for each mark of ammunition.

21. *Practice indirect ball-firing with spirit level.*

One way of doing this is to put correct elevation on the tangent sight, and place the bottom of a straight-edge from foresight to tangent sight slide, with a spirit level on the top edge. Elevate the gun until the straight-edge is level. The laying for direction is given by compass reading. *In all indirect ball-firing* there is much data which has to be found out on each occasion—namely, (1) distance of target above or below axis of barrel; (2) height of ground above axis of barrel between gun and target; (3) horizontal range of summit of intervening ground from gun; (4) probable distance below the given trajectory and the lowest bullet in cone of fire: this must clear the intervening ground, or gun site must be moved farther away or nearer to the hill according to trajectory of the range; (5) if on far side of hill, the bullets go over one's own troops. The distance between lowest bullet in cone of fire and top of one's own trenches must not be less than that laid down as safety angle.

22. *Practice indirect ball-firing with Abney level.*

The same remarks apply here as in No. 21, except that the required angle of elevation is first set on the Abney level, which is then placed on the cover with eye-piece to rear and gun elevated until the bubble is centred.

23. *Practice overhead ball-firing by tangent sight method.*

This is one kind of direct fire which can be used against infantry advancing or retiring in full view. In other than trench tactics this kind of sighting could well be very useful.

24. *Practice overhead ball-firing by safety angle method.*

The factors concerning this method are few in number, but they have to be worked out for each range separately. This might better be called "safety angle factor," as it is frequently employed in direct and indirect fire; in these two kinds the guns may have to fire over their own infantry, which may in turn be stationary in trenches or advancing on the enemy. In both cases, when first opening fire the problem of a safe clearance of the infantry is the same. It must be found out what the relationship of the infantry's position is to the axis of the machine-gun barrel, when laid horizontal. When the infantry advance, another factor comes in, but this is also called the "safety angle"; and when our troops enter this angle all the guns in the formation must be given more elevation, so as to lift the fire, and this lifting must be continued as long as our troops are advancing. Now, this "lifting the fire" requires a very careful training by means of drill and detail, carried out on aiming targets before going to the range. The lifting of direct firing should be carried out first, and very large elementary targets should be used. When practising safety angle for troops in trenches, a large square screen should be placed on the parapet of the trench, so as to show what the distance is of the lowest bullet from centre of bullet sheaf.

25. *Drawing attack and defence range cards.*

This work comes in the training of that specialist, the range taker and ground scout. The objects selected should vary in range, so that they in turn can be used as group keys. Much training in judgment will be required to enable the scout to select such points as the enemy now occupies or will occupy in the advance or retirement. In the preliminary training good judgment and great accuracy must be considered before the length of time taken. But in advanced training, cover from view, quickness of judgment, reading instrument and drawing in card must be insisted on.

26. *Physical training in carrying machine gun, tripod, or two full belt-boxes.*

A man must be able to double 600 yards with any of the above, and go into action at once. Begin with 100 yards and gradually increase.

Machine gunners must always be in good physical training and strong enough to carry any part of the equipment at the double up to 600 yards. They must keep up their physical hardness all the time of their service.

SUGGESTIONS AND NOTES ON THE STUDY OF GROUND

By Major F. V. Longstaff

The reading of ground is absolutely essential to the making of a good machine gunner.

The following points are noted in the hope that they may be of some help to the machine-gun officer who is anxious to acquire some knowledge of this subject. The qualities necessary to success in the reading of ground do not come to all by instinct, but an officer, by taking pains, may acquire them.

"Du terrain, encore du terrain, et toujours du terrain." We must never cease to look at it, and think of it, until we can read it as easily as the biggest poster, and as instinctively as we read "Lipton's Tea," when it flashes out in huge, many-coloured letters from across the Thames. "Ground," too, to us, should be more interesting, and appear better printed than the best book we have ever read. For it is *the* book of books for us, simply because it is on its ever-varying pages that we have to play the game of our lives. We must be reading *our* book *always;* when we walk, when we ride, when we travel in the train, when we hunt, when we deer-stalk, when we shoot, when we fish—in a word, whenever we do anything which causes our eyes to rest on its pages. Again, as the beautifully varying pages present themselves we should set ourselves imaginary problems on them, and find a solution for each. We must make this habit second nature to us; then, when we are suddenly confronted with a difficult problem in the heat and anxiety of a *real* examination—*i.e.,* on active service—we shall be like the mathematician who has all the necessary formulæ at his finger-ends, and we shall find a solution to the problem.*

So writes General Alderson in a striking passage in his "One Hundred Notes." This practised leader of men is telling us how he has learned his own lesson. It is a remarkable fact that many of our great Generals learnt the art of ground reading in India, Africa, or North America, where there are wide spaces and all kinds of ground covering and land forms. It may also begin by one being dropped in a country of long distances, and then having to make oneself feel at home there. It is important not to become dependent on such a large-scale map as the one-inch, and the reading of all kinds of maps should be much practised. Officers should not blame the ground if it does not agree with the map.

When advancing in a strange country, care should be taken not to be let down by the map, using one's eyes in reading the ground, so as to act as a check to the map.

One good way is to take an officer out for individual instruction. When an instructor meets a pupil for the first time, he must find out what habits of action or thought the

* Alderson, p. 39.

pupil has that can be used as a foundation upon which to build the new knowledge and habit. The following method of learning how to read ground is that which can be employed by an instructor giving individual or collective training:

Most officers have visited some spot in the country year after year, and from that spot there may be one view strongly fixed in their memory. This is a good foundation on which to build a capacity for reading ground. The best start is to sketch that view on paper and insert all details that can be remembered —trees, rocks, fences, gates, haystacks, cottages, paths, and the differences in visibility between early morning, noon, and evening. Try and remember another view looking up a hill near the same locality; also one looking down from the above hill in the reverse direction. It is absolutely necessary after studying a cross-country view to a certain spot (A), while at (B), to be able to visualize the view of (B) in the reverse direction, and decide whether you would be visible at (B) from (A). This should be practised both standing and prone.

The next step is to increase one's descriptive vocabulary, both of what objects look like under all conditions and the proper descriptive name for each. Objects one sees, and should know the descriptive name for, may be roughly classed under four heads:

(i.) Water, including seas, lakes, ponds, rivers and other streams, canals, swamps, etc.

(ii.) Relief, including mountains, hills, valleys, cliffs, etc.

(iii.) Culture—*i.e.*, works of man, such as towns, cities, roads, railways, boundaries, buildings, crops, etc.

(iv.) Living creatures—man, domesticated and wild animals, etc.

Daily training in observation, by noting and remembering small signs or details as one walks along the road or street, is important. One good exercise is to count the number of windows in the front of a house, and then the number of separate panes of glass. By frequent practice one will grasp the peculiarity of each house front at a glance, and so build up a unit of measurement, for use on solitary country houses or cottages, on ground one is reading for tactical employment. One should make measured drawings of several sizes of windows, so as to be able to judge correctly the height, etc., of a distant house which in turn can be used as a unit of measurement to get the height of a hill, etc.

The use of shadow is of great assistance in the visual reading of ground. There are two kinds: solid shadows made by the sun on opaque substances (cliffs, trees, or buildings), and transparent shadows made by the sun through clouds which are moving.

In new countries streams and rivers are looked upon by travellers as guides and friends. For several years in a mountainous newly-settled country, the author could only obtain maps showing the rivers, the altitudes of some mountain summits and points on rivers, no contouring lines being given. In a very short time after studying the elementary geology of the country, he was able to read the mountains and hills into the maps, using the rivers as his framework. In many other parts of the world he has employed the rivers and their watersheds to enable him to visualize the whole area bounded by the water-partings.

If we return to our well-known spot in the country—there is probably either a stream or a large ditch which joins a stream; follow this mentally all the way till it joins the main river which flows into the sea. Try and visualize the whole journey of the water in our little stream till it reaches the sea. Make a sketch of it, and note how many feet it is above the sea at our spot; we then know its fall to the sea, but this fall is not regular as a rule.

On opening a new map make sure that the north is on the top edge. Find the largest river or stream which runs off the sheet, and follow it upstream, noting the tributaries entering it from either bank, and so up to its source on the water-parting on the edge of the watershed. Perhaps there is a tributary stream of the same watershed which rises on the other side of the water-parting or divide, in which case the water-parting is not the edge of the main watershed. From the source return down the stream, and find out the number of feet it falls in each mile on its way to the main river and the sea. Ascend the stream again to within a few miles of its source, run an offset at about right angles to its course, and measure how many miles to the next stream and how high above sea-level is the intervening ground. If the map has contours, make a section with the perpendicular scale ten times that of the horizontal.

Draw out to the same scale as above some trajectories between 1,500 and 2,800 yards, on tracing cloth. Place the tracing cloth over the map section and see which trajectories will clear the map section, if any.

The reading of ground for use in fire tactics is largely a question of (i.) height of summit above surrounding country; (ii.) difference in level between object and possible machine-gun site; (iii.) shape of ground between object and machine-gun site.

Rivers and streams can be made the best guide to the reading of main ground features all over the world. In those countries where rivers dry up in the hot season it is, of course, the river valleys which are the guide, while the direction of flow can be seen in the dry sand of the river-bed.

There are some excellent notes on reading maps in "The Interpretation of Topographic Maps," Washington, U.S.A., 1908, Department of the Interior. A few are given here: "All water features are shown in blue, the smaller streams and canals in full blue lines, and the larger streams, lakes, and the sea, by water lining. Certain streams, however, which flow only a part of the year—their beds being dry at other times—are shown, not by full lines, but by lines of dots and dashes. Ponds which are dry during a part of the year are shown by oblique parallel lines. Relief is shown by contour lines in brown. Each contour passes through points which have the same altitude. A succession of these contour lines far apart on the map indicates a gentle slope; if close together, a steep slope; and if the contours run together in one line, as if each were vertically under the one above it, they indicate a cliff." Of course, the "conventional signs" of the map in use must be substituted, where necessary, in the above extract.

Since rivers and streams form such a helpful guide to the reading of new ground, it is worth while to give them some study. This subject comes under the heading "River Valleys," in the "Physical Geography,"* from which the following is an extract:

"A river is a natural drainage line on the land, and it usually occupies a valley which has certain quite definite characteristics. On either side of the river are the *valley sides*, or walls, sometimes rising gently, sometimes steeply: at times to great heights, and, again, with only low elevations. In the lowest part of the valley, and generally near its middle portion, the river flows, usually with a meandering course. In most cases the river is immediately bounded by rather steeply rising *banks*, only a few feet in elevation, beyond which the slope is more gentle, and sometimes even a plain, which in times of flood is covered with water. Here and there, at irregular intervals, *tributaries* enter the main stream, and these themselves branch into other tributaries, until very often there is finally produced a branching network of minor streams, all directly or indirectly contributing to the supply of water in the main stream. This principal supply often comes from springs, and sometimes the source of the river is a spring. Together these form a *river system*: and such a system may have an area of only a few square miles, or it may drain an area of many thousand square miles. The line or plane which separates one stream or system of streams from another is known as a *divide*, or *water-parting*.

"All streams have these general characteristics; but when their valleys are examined in detail, there are found to be many differences, not merely between different rivers, but even in the several parts of the same river. The valley of one stream, or part of a stream, may be a pre-

* "Elementary Physical Geography," chap. xv., by R. S. Tarr, B.S., F.G.S.A., New York. Macmillan, 1895.

cipitous gorge, or it may have very gently sloping sides. In some rivers there are flood-plains and deltas, in others these are absent; and in some cases the rock walls of the valley may rise directly out of the river. In some there is a permanent flow of water, in others the supply is intermittent, and in some extreme cases water flows only once or twice a year. In most river systems the tributaries are numerous, but in some cases they are few; and while some of these join the main streams at a high angle, in many cases they enter it at an acute angle. There are many reasons for these differences in streams, the two most important being the position and kind of rock in which they occur, and the stage in development which they have reached.

" We must bear in mind that the river valley is the result of the combined action of stream erosion, which tends to deepen, weathering, which tends to broaden the valleys, and the transportation of sediment furnished by these means. Erosion proceeds more rapidly than weathering, but there comes a time when its action is checked. In no part of the valley can the stream cut below the sea-level, or below *base-level,* as it is called; and, since it must carry water down a slope, in its erosion the river reaches lower levels near its mouth than higher up in its course. Until a line of easy slope is reached, erosion exceeds weathering; but then, since erosion is checked while weathering continues, the latter produces its most marked effect, and the valley gradually broadens, while the hills slowly melt down. Unless interfered with, this would continue until the surface was *base-levelled,* or reduced to a nearly level condition.

" The development of the stream proceeds most rapidly near its mouth, and later in the head-waters; and, consequently, tributaries are not numerous at first. But one by one they begin to develop, until all the area is brought under the influence of some stream or rill. At first the divides are not very definite, and they may be flat-topped and swampy; but in maturity these become quite sharply defined, and usually every part of the area is drained.

" At first the topography *guides* the stream course, but finally the river course *determines* the topography.

" Between any two streams there is a line, or an area, which divides the waters, sending a part one way and the rest in an opposite direction. These divides, or water-partings, are by no means permanent, but are constantly and usually very slowly changing. The stream that has the most power pushes the divide into the territory of the other, and there are various ways in which one stream may have more power than another. One may have a shorter course to the same level, and hence have a greater slope; or one may be cutting through soft rock while the opponent is working in hard layers; or the rainfall on one side of the divide may exceed that on the other. Gradually the divide moves into the area of the stream having the least rainfall, or the least slope, or the hardest rock."

The reading of ground is largely a question of habit and practice. When viewing new ground most people automatically compare it with some well-known scenery. It is most important for machine-gun officers to endeavour to improve and strengthen this habit. They should note the number of objects, such as stream, ditch, road, footpath and undulation of the ground; count every tree, post, house, window, and door. They should estimate the range of each group of objects, for it will generally be found that these naturally form themselves into groups large or small, and the combination of ranges and groupings will subdivide many of these into different planes, because of the different ranges of objects in the same group. This last property can be easily seen in a roadstead full of vessels at anchor, which, when seen from the shore, appear to be all in a line, but on going out to them some are found to be much farther off than others. On returning to the same point on shore, though the same view is again seen, yet somehow it is different because we have been out there, and know. This is what is meant by dividing scenery up into planes. This habit of seeing in planes requires most careful cultivation, and in the case of most good ground readers it has taken many years of practice to acquire. Another good exercise can be got from photography. A view is selected by the eye; it is found in the view-finder more or less, but the negative gives a scene not at all like what the eye really saw. The chief reason is that the camera was held at chest height instead of at eye height; of course a different kind of view-finder is necessary if the camera is to be held at the eye height. But the question of the height of the eye above the ground, when reading it for military purposes, is very important, and apt to be forgotten at first; all ground should be read lying, and not riding or standing.

The habit of *seeing in planes and counting all objects seen without moving one's head* cannot be practised too often, and it can be practised on any ground however familiar. It is only after one has thoroughly read, reread, and well thought out the peculiarities of a well-known piece of ground, that the next step should be taken of attempting new ground and scenery. Suppose this ground is a mile square, with its four sides facing the four points of the compass. Remember that this ground can be seen over in four main different directions, north, south, east, and west. From each direction the same view will vary with the time of day from the sun's position, the presence or absence of clouds, and the clearness of the atmosphere. The seasons of the year will govern the presence or absence of foliage, and this will affect the range of view. Most views in the field are composed of foreground, middle distance, and background. In views where there is a lot of blue in the atmosphere a pair of yellow-tinted spectacles will remove this in a wonderful manner. The yellow glass should be quite flat, as it simply acts as a light filter, and may also be made to fit in front of the object glass of field-glasses or telescopes. This yellow glass will enable one to look towards the sun and discern objects without straining the eyes.

In many areas of undulating country the main ridges all run in the same direction, and while one side is steep the other side is a gentle slope. This is where a sound knowledge of geology comes in most usefully, as by getting up the directions of the folds of the earth's surface of that part of the country where one is going on active service, a foundation can be laid in one's mind of the area of ground of which one will have to read a part. This will often enable one to be sure what the shape of the far side of a hill or ridge is without seeing it. A common mistake is to read an end-on view of a ridge of hills into an isolated hill.

Most people stand square with the object they are looking

at; but when reading ground for use with fire tactics, one should form the habit of looking half right and half left, because oblique fire is one of the principles of machine-gun tactics. The habit of *seeing in planes* should also be employed in *oblique* ground reading, but in this case they should follow the main direction of the enemy's line, rather than be at right angles to one's own line of sight.

A knowledge of local habits and conditions should enable one to forecast the presence or absence of buildings or other features out of the line of sight. It is generally possible to get hold of the chief peculiarities of the ground before going to the scene of war.

As regards Europe and many parts of Asia, the usual cockpits are well known to the student of military history. He has tabulated the number of battles which have taken place in each area or cockpit, the date, number on each side, the moral of each side and the way the tactics were applied to the ground. The General Staff of each nation with a modern army should have "a worked-out scheme" how to employ, to its own advantage, the main land features in whatever part of the world it may have to fight. The General Staff of a nation should be the brain, and is, in a way, like a consultant in medicine, engineering, law, architecture, chemistry or accountancy. It has also to collect information of all kinds which it has to be always sorting, digesting, and recording in a suitable form for further use; after which parts of it are used in working out and keeping "up-to-date" schemes similar to the foregoing. In the British Empire the Staff College at Camberley trains officers for the General Staff. In some European countries the corresponding educational establishment is called the War School or College—that is to say, the place where the highest phase of the art of war is taught. In the entrance examination for the Staff College, under the subject of "Military Geography," is found in the syllabus:

Europe generally, with special reference to the frontiers of countries, physical features affecting strategical considerations and military operations. Fortresses and fortified places; main lines of communications by land and sea; coasts, harbours, and coaling-stations. India, Burmah, and all British colonial possessions, considered in the above respects. Persia, Central Asia, and the Far East generally, as above. North America, a general knowledge only. Egypt and the Nile Valley. The Red Sea littoral and Somaliland. The principal trade-routes of the world; naval bases; important ports, coaling-stations and cables. Questions on any geographical subject which is occupying general attention.

Note.—Special attention should be directed to important frontiers of British possessions and the countries adjacent to them. Without expecting great detail, candidates will be required to have a general knowledge of the chief characteristics of the people of the countries included in, and adjacent to, the British Empire; also of climatic conditions and seasons, in so far as they are likely to affect military operations.

Remember the above syllabus only covers what an officer has to pass in to be qualified for admittance, and that it is only the foundation upon which to build a more extended teaching of military history and geography, strategy and tactics.

From this it is clear that the study of ground must not be confined to any particular country or continent. There are many parts of the world, where troops are now fighting, where the scale of the land-forms is much vaster than on the Western front. In the mountainous parts of the world, it is sometimes very difficult when travelling in the valleys to take in the true scale. When marching, one finds that a large shoulder of a mountain, which appeared quite near in the morning, is apparently just as far away at midday. One deceptive kind of land form is when the whole shape is in such excellent proportion that its huge size is not at once realized. It often requires a very strong glass to spot a unit of measure (such as an animal, tree or house) which at once sends the land form in question either flying away to the sky-line or travelling rapidly closer in towards one. Officers must, when first landing in a strange country, build up a new set of eye-measuring standards based on distances measured from an accurate map.

The reading of mountains of all kinds—bare rock, grass covered, tree covered, or snow covered all the year—is a very large and difficult subject on which a whole book could well be written. But try our old friend the stream, and see what we can learn from it, to enable us to read that wonderful country far above the tree limit. (It is a simple book to read if one can understand the quality of snow, ice, and rocks.) We should try to understand the annual life-cycle of the high mountain flora, together with the height and density of the underbrush and bushes, found in the avalanche tracks between the trees and in the open.

Looking up the side of a mountain range from a valley, streams are seen falling down channels and over cliffs into the valley, where they join the main river torrent. Apply here the habit of *seeing in planes*, and make an estimate of the heights of the cliffs and pieces of rocks. After well marking them down, climb up to them and check your estimates. Probably the first cliff will stop your ascent, and you will have to move to the right or left to find a way up the mountain. Return to your observation post in the valley and look up the side of the mountain range again. How very different it looks now, and how well your legs and body know the heights of those small bits of cliffs and stones! An ordinary deer-stalking telescope is the best auxiliary aid to reading mountains. It should be used with your back against sloping ground or a rock, your left wrist resting on your knees holding the glass. Remember it is much easier to find a way up a mountain than down. At the same time the range of 2,800 yards should enable machine-gun companies to apply their fire all over the enemy without much or any changing of position. Small terraces or valleys can often be found from which indirect fire can be used and so be hidden from the artillery. A clinometer, with telescope and cross hairs, is necessary in the mountains to find out whether you are above or below your target.

The best official thoughts on the "reading of ground" are to be found in "Field Artillery Training, 1914," chap. xii., Reconnaissance Duties, etc. The subject is dealt with very concisely and clearly, and in every way the chapter is most valuable for the instructor. Two extracts are given as examples:

> The primary object of artillery reconnaissance is to collect and transmit information, which is important from an artillery point of view, to the commander who orders the reconnaissance. At the same time anything which is likely to prove of value to the other arms should be noted. While smokeless powder and long-range weapons have increased the difficulties of reconnaissance, they have also added to the value and importance of accurate information. It is essential, therefore, that artillery should be well trained in this difficult work.
>
> In each artillery brigade there should be at least one officer's patrol, the members of which are accustomed to work together, and capable of carrying out a combined reconnaissance.

> The men should be taken to some ground with a good view all round, and each in turn questioned as to what he makes of distant objects, such as men, cattle, bushes, rocks, etc., his answers being checked with field-glasses by the instructor.

APPENDICES

ABBREVIATIONS IN APPENDIX I

Abbreviation.	Name of Newspaper or Periodical.		Place of Publication.
A.M.B. ..	Artilleristische Monatshefte ..	..	Berlin.
A.M.S. ..	Proceedings of the Aldershot Military Society	..	London.
B.P.B.M.	Bulletin de la Presse et de la Bibliographie Militaires (Supplement to J.M.O.B.) ..	F.	Brussels.
C.J. ..	Cavalry Journal	Q.	London.
D.M.S. ..	De Militaire Spectator	M.	Haarlem.
I.R. ..	Internationale Revue (Armee und Flotten)..	M.	Dresden.
J.D.A.M.	Jahrbücher für die Deutsche Armee und Marine	M.	Berlin.
J.M.O.B.	Journal Militaire Officiel ..	M.	Brussels
J.M.S.I. ..	Journal of the Military Service Institution	2 M.	New York.
J.M.S.S. ..	Journal of the Military Scientific Society	Q.	Petrograd.
J.S.M. ..	Journal des Sciences Militaires ..	½ M.	Paris.
J.U.S.I. ..	Journal of the Royal United Service Institution	M.	London.
K.M. ..	Kavalleristische Monatshefte ..	..	Berlin.
K.T.Z. ..	Kriegstechnische Zeitschrift ..	..	Berlin.
L.R.I. ..	La Revue d'Infanterie	M.	Paris.
M.A.G. ..	Mittn. über Gegenstände des Art.- u. Genie-Wesens ..	M.	Vienna.
M.W.B. ..	Militär-Wochenblatt	⅓ W.	Berlin.
O.M.Z. ..	Streffeurs militarische Zeitschrift, zugleich Organ der militar-wissenschaftlichen Vereine ..	M.	Vienna.
P.R.A.I. ..	Journal of the Royal Artillery ..	M.	Woolwich.
P.U.S.I. ..	Journal of the United Service Institution of India	Q.	Simla.
R.A.G. ..	Rivista di Artigliera e Genio ..	M.	Rome.
R.C. ..	Revue de Cavalerie	M.	Paris.
R.d'A. ..	Revue d'Artillerie	M.	Paris.
R.M.B. ..	Revue de l'Armée Belge ..	2 M.	Liège.
R.M.S. ..	Revue Militaire Suisse	M.	Lucerne.
R.P.M.I.	Recent Publications of Military Interest. War Office ..	Q.	London.
S.Z.A.G. ..	Schweizerische Zeitschrift für Artillerie und Genie	M.	Frauenfeld, Switzerland.
U.S.M. ..	United Service Magazine ..	M.	London.
V.S. ..	Voyennii Sbornik	M.	Petrograd.

APPENDIX I

A BIBLIOGRAPHY OF UNOFFICIAL BOOKS, PAMPHLETS, AND ARTICLES CONCERNING THE SUBJECT OF MACHINE GUNS

THE following Bibliography does not claim to be exhaustive, but is believed to be fairly complete. It is based upon research in the library of the Royal United Service Institution, with additions from the notes on books in "Recent Publications of Military Interest," and on the numerous notices of books on the subject contained in Commandant Lavau's work. It represents the labour of many months in collecting and arranging the various entries.

The Bibliography has been arranged in order of time, and thus forms a record of the gradual development of the study of Machine Guns and of the literature of the subject.

The authors will be glad to receive additions duly authenticated to be inserted in future editions.

1862.

VANDENBURGH, GENERAL. (U.S.)

"A New System of Artillery for Projecting a Group or Cluster of Shot." *J.U.S.I.*, vol. vi., p. 377. Lecture.

1868.

TACKELS, CAPTAIN C. J. (French.)

"Armes de Guerre." (*Note.*—This is a good textbook of firearms with many plates.—F. V. L.)

1869.

CLAXTON, COLONEL. (Belgian.)

"Les Mitrailleuses, ou Armes Mécaniques." Liège. (*Note.*—This gives good illustrations of field mountings.—F. V. L.)

FOSBERY, MAJOR G. V., V.C. (British.)

"Mitrailleurs and their Place in Wars of the Future." *J.U.S.I.* pp. 539-563. Illustration of the "Montigny" mounted on 6-pr. service carriage. See also the Rotunda at Woolwich. (*Note.*—This covers that period of plenteous inventions which included the acceptance of the breech-loading rifle and the metal cartridge case. The great weight of the early mitrailleurs, the imperfections of the metal cartridge case, the thick smoke of the powder and the claimed similarity of the bullet sheaf to the shrapnel sheaf, were the factors which prevented the sound tactical position of mitrailleurs being determined. —F. V. L.)

1870.

GATLING, DR. (U.S.)

"The Gatling Gun." *J.U.S.I.*, vol. xiv., pp. 504–528. 4 illustrations. Contributed.

1871.

WILLE, LIEUT. R. (Prussian.)

"On Case Shot Guns." Berlin.

1872.

FLETCHER, COLONEL. (British.)

"Employment of Mitrailleurs." *J.U.S.I.*, vol. xvi., pp. 28–55. 1 illustration. Lecture on January 22, 1872. Chairman, Major-General F. Eardley-Wilmot, R.A.

1873.

FOLGER, LIEUT.-COMMANDER W. M. (U.S. Navy.)

"Adaptability of the Gatling Gun for Naval Purposes."

FOLGER, LIEUT.-COMMANDER W. M. (U.S. Navy.)

"The French Mitrailleuse: a Full and Complete Description of its Construction, Service, etc." Translated from the German. Washington: Government Printing Office. 12 pp. and 4 folding plates. (*Note.*—This is one of the few complete descriptions of the original Mitrailleuse of 1870. The great difficulty of reliable details is well expressed *at the end of the preface:* "After the capture of the first specimens, and their exhibition at several points in Germany, descriptions and sketches in explanation of the principles of construction, method of service, etc., appeared in the current journals; but as the French in deserting the guns had usually rendered them useless by destroying or carrying off some portions, these descriptions were frequently incomplete and faulty. In the following paper an endeavour is made to render plain all details."—F. V. L.)

1874.

FRANKLIN, MAJOR-GENERAL, W. B. (U.S. Army.)

" The Gatling Gun." (*Note.*—History of its introduction.—F. V. L.)

ORMESSON, De. (French.)

" On the Tactical Employment of Field Artillery and Mitrailleuses in Connection with the Other Arms." Paris, Canera.

OWEN, CAPTAIN J. F., R.A. (British.)

" Compound Guns, Many-Barrelled Rifle Batteries, Machine Guns, or Mitrailleurs." Mitchell and Co., London. 38 pp., 3 folding plates: 0·45-inch Service Gatling, 10-barrelled Palmcrantz, Lock of Gatling. (*Note.*—This is the best work I know of at that time on Machine Guns. It is a first-class compilation on the subject, describing the different systems of guns, what European countries were doing on the subject, what experiments were carried out, both in Great Britain and on the Continent. He deals with the tactics, giving the opinions of well-known soldiers of that period and his own conclusions. Colonel Wray's committee, in 1870, carried out a series of firing experiments with small Gatling, Montigny Mitrailleur, 12-pr. B.L., and 9-pr. M.L. Shrapnel shell only used with the field guns. The ranges were 300, 400, 600, 800, and 1,000 yards, and the total hits secured were: Small Gatling 2,803, Montigny 1,708, 12-pr. B.L. 2,286, 9-pr. M.L. 2,207. The committee recommended that Gatling Batteries be formed of 12 guns, 5 officers, 101 N.C.O.'s and men, 6 small-arm ammunition carts, 6,708 cartridges per gun, 34 riding horses, and 56 draught horses. The road-space occupied by a battery of 6 9-pr. M.L. Field Guns is 353 yards, while that of the battery of 12 Gatlings is only 156 yards. On p. 10 Colonel Fielding is quoted: " The proper use of Mitrailleurs to be as representing a certain number of infantry, for which there is not room on the ground, suddenly placed forward at the proper moment, at a decisive point, to bring a crushing fire on the enemy." There is such important historical information in the Preface by Captain Owen that it is given here in full. " Mitrailleurs or Machine Guns have attracted a great deal of attention of late years, and now constitute a part of the armament of most great Powers. A succinct account of their gradual introduction, and description of those used by different nations, will therefore, I trust, be found of some interest. In 1869 our own Government, finding that in France and Belgium weapons of this nature were being manufactured in considerable numbers, purchased Mitrailleurs, both on the Montigny and Gatling principles; and in 1870 a Special Committee was appointed to report

upon them, and on Mitrailleurs in general. To the two reports of this Committee, dated November, 1870, and November, 1871; to an interesting lecture delivered at the U.S. Institution, London, in January, 1872, by Colonel Fletcher, Scots Fusilier Guards (one of the Special Committee); and also to the report of a Swedo-Norwegian Commission of Officers (see *P.R.A.I.* before 1874), and an article upon the same in the *R.d'A.*, February, 1874, I am indebted for most of the information given in the following pamphlet. Dated July, 1874."
—F. V. L.)

1875.

MARVIN, COMMANDER J. D. (U.S. Navy.)

"Gatling Guns, Instructions for Use and Care of." Published by U.S. Navy Bureau of Ordnance. 43 pp., 6 folding plates. The guns are the model of 1874, long and short. The former has the gravity drum feed, the latter the gravity slide feed. Full details are given of the drill, mechanism, and care of the equipment, with a few pages of its minor tactics on sea and land. The guns are sighted up to 1,000 yards. An extract is given from p. 20: "Manual of the Long Gatling Gun, mounted on the field carriage, and equipped for service on shore, using feed drums.—This drill presupposes the ship to be supplied with two guns and four drums, and that each gun-carriage is fitted to carry, besides a spare article box under the axle, either the gun and two ammunition boxes, or, if the gun be dismounted, three ammunition boxes, and that in this service one carriage is used as a limber, and is attached to the gun-carriage proper, as shown in Plate I. The gun's crew consists of 13 men—No. 1, gun captain; 2, 1st loader; 3, 1st crankman; 4, 2nd loader; 5, 2nd crankman and 3rd loader; 6, 7, 8, 9, fillers; 10, opener; 11, 12, supernumeraries; 13, artificer." On p. 37 it says ": . . . Practice at the rate of 480 per minute has been made with the Experiment Battery . . ."; p. 41: "Beyond 800 yards the accuracy of the short gun, calibre ·50, cannot be implicitly relied upon. The larger guns of the same calibre will be found perfectly reliable at 1,500 yards."

ROGERS, CAPTAIN E. (British.)

"The Gatling Gun: its Place in Tactics." *J.U.S.I.*, No. 32, pp. 419–445. Two illustrations of a Gatling gun on a cart. Colonel H. C. Fletcher in the Chair; and the following officers took part in the discussion:—Captain J. F. Owen, Major Hale, General Sir R. P. Douglas, Bart., Captain Nolan, M.P., Admiral Sir H. Codrington, K.C.B., Captain Tulloch. (*Note.*—This is a valuable review of data and opinions up to date on the subject of the lecture. It urges that Gatlingeers be thoroughly trained in the use of the machine, that the Gatling be not confounded with artillery, and that no foreign army will in future take the field unprovided with a complement of Machine Guns, and we must be prepared to meet them with at least equal weapons and superior tactics.—F. V. L.)

1878.

JAMES, LIEUT. W. H. (British.)

"Repeating Rifles." *J.U.S.I.*, No. 98, pp. 1090–1102. Three tables. (*Note.*—This is a short account of the Repeating Rifles in use by France, Norway, Austria and Switzerland; the Hotchkiss, the Kropatschek and the Krag. This was an article.—F. V. L.)

1879.

BAINBRIDGE, CAPTAIN E., R.A. (British.)

"Practice with the 1-in. Nordenfelt Machine Gun at Buckan, December 22, 1879. *P.R.A.I.*, 4 pp.

"Description of 1-in. Torpedo Battery Gun," 3 pp. *P.R.A.I.*

FOLGER, W. M. (United States.)

"Adaptability of the Gatling Gun for Naval Purposes," 16 pp.

1880.

GAZETTE, ARMY AND NAVY. (British.)

From this year onwards the *Army and Navy Gazette* has notices and articles on Machine Guns.

JAMES, CAPTAIN W. H. (British.)

"Modern Fire: its Influence on Armament, Training and Tactics." *J.U.S.I.*, vol. xxiv., pp. 378–403. Lecture, May 7, 1880. Chairman, Lieut.-General C. P. B. Walker, C.B. The following officers took part in the discussion:—Major T. Fraser, Colonel C. B. Brackenbury, Captain G. W. Cockburn, Lieut.-Colonel Lonsdale Hale, Mr. C. F. Lowe, and Sir W. Codrington. (*Note.*—This contains a number of sound and clear statements on fire tactics, some of them quite prophetic. The lecturer is a little ahead of his time, but he deals with the problems in large European Wars. I give extracts:—Deadliness of fire can be increased (1) by increasing the flatness of the trajectory; (2) by increasing the number of bullets poured on a certain spot. The number of bullets fired at a given object may be increased (1) by firing at longer ranges; (2) by firing more often, the latter by the use of repeating rifles or other means of increasing rapidity of fire. There are many useful tables on musketry results.—F. V. L.)

NORDENFELT, T., Esq. (Swedish.)

"The Nordenfelt Machine Guns." *J.U.S.I.*, No. 108, pp. 785–805. 1 plate and 3 tables. Admiral Sir F. W. E. Nicholson, Bart., C.B., in the Chair. The following officers took part in the discussion :—

Major E. Rogers, Commander Curtis, R.N., Admiral Selwyn, Major L'Amy. (*Note.*—In Sweden the Nordenfelt Machine Gun is known as Palmcrantz's System. This is a clear description of the Nordenfelt system and the variations in calibres and number of barrels. The use both on war ships, in forts and in the field is roughly indicated. Tactics are not dealt with to any extent. Many experiments in ball-firing are quoted. Major E. Rogers, who took a large part in the discussion, expressed surprise at the little progress made in *organization and tactics* since 1875.—F. V. L.)

1881.

GREENER, W. W. (British.)

" The Gun and its Development, with Notes on Shooting." London, Cassell. 674 pp., two indexes, and many good illustrations. See p. 291 for history, evolution, and manufacture of cartridges; pp. 467-471 for Machine Guns.

TRIALS. (British.)

" Machine Guns." *The Volunteer Review,* January 22, 1881. Editorial. " Machine Guns at Shoeburyness." *The Volunteer Review,* January 22, 1881. Article.

TURNER, CAPTAIN A. E., R.H.A. (British.)

" Trials of Nordenfelt and Hotchkiss Machine Guns in Many Countries." 4 pp. From the *Revue d'Artillerie,* February, 1881.

1882.

FOSBERY, LIEUT.-COLONEL G. V., V.C. (British.)

" Magazine Rifles." *J.U.S.I.,* 1882, No. cxvi., pp. 456–484. One plate of the following rifles:—Martini, Enfield, Comblain, Burton, and Deeley-Edge. Lecture May 12, 1882. Chairman, Colonel Lord Elcho, A.D.C., M.P. The following officers took part in the discussion:—Captain W. Arthur, C.B., R.N., Admiral H. Boys, Admiral Selwyn, Sir L. Graham, and Lieut.-Colonel Hope, V.C. (*Note.*—This deals with the introduction of magazine rifles, the consequent arguments of shortage in ammunition supply, and difficulty of fire control. It is most instructive for a student. The following extract is instructive:—" A most ably conducted and exhaustive set of trials has already determined that we have a system of Machine Guns by which one or two men can do as much destruction as, say, forty ordinary soldiers, and that without the chance of deranged mechanism or the complications which formerly militated against their use; and yet, though it is certain that an enemy will always do the unexpected thing, and if an European one, use Machine Guns against us, we neglect to acquire them for the Army because the exact tactical place for the weapon is as yet undiscovered."—F. V. L.)

GARDNER, MR. W. (U.S.)

"Machine Guns and How to Use Them." *J.U.S.I.*, 1882, No. cxiv., pp. 103–114. One plate, showing Gardner guns of 1, 2 and 5 barrels, their mechanism, mountings and transport. Lecture, February 24, 1882. Chairman, Vice-Admiral H. Boys. The following officers took part in the discussion:—Major-General Sir E. Hamley, Lieut.-General Sir B. Walker, Major-General Lord Chelmsford, and Captain Cleveland, R.N. (*Note.*—This is largely on the mechanism and mounting of the Gardner gun. However, Mr. Gardner (late Captain U.S. Army) gives many sound tactical ideas, of which a few are given. "Arms (Machine Guns) should be provided in far greater quantity than is barely sufficient to supply the number of soldiers at the disposal of any country." "I believe most of the solid work of war is now done in the trenches, in their making and defence." "How fast can you shoot?" "I make a distinction between a cartridge destroying machine and a Machine Gun. It is the number of hits on the enemy, not the number of shots fired, that we care to score." "I should place the qualities of a Machine Gun in order of merit about as follows: (*a*) Reliability; (*b*) Strength; (*c*) Weight; (*d*) Simplicity; (*e*) Durability; (*f*) Ease of manipulation; (*g*) Rapidity of fire."—F. V. L.)

NAVY. (British.)

"Machine Guns in the Navy." From *The British Navy: its Strength, Resources, and Administration*, by Sir T. Brassey, K.C.B., M.P., M.A., vol. ii. Miscellaneous subjects connected with shipbuilding for the purposes of war. London: Longmans, Green and Co., 1882. In Chapter II.—Guns and Gunnery—pp. 78-81, Machine Guns are dealt with.

1883.

BERESFORD, CAPTAIN LORD CHARLES, R.N. (British.)

"Machine Guns." *J.U.S.I.*, 1883, No. cxxi., pp. 602–628. 11 tables, 3 plates showing Nordenfelt quick-firing gun on land and sea-mountings, Hotchkiss revolving cannon on sea mounting, Hotchkiss 1-pounder quick-firing on sea mounting, Gatling gun 45-inch on land (wheel) mounting. Lecture June 15, 1883. Chairman, Admiral Sir F. W. E. Nicolson, Bart., C.B. The following officers took part in the discussion:—Vice-Admiral Vesey Hamilton, C.B., Captain P. H. Colomb, R.N., Vice-Admiral H. Boys, Major Rogers, and Mr. J. B. Fenby. (*Note.*—Most of this lecture is devoted to Machine Guns as applicable to the Naval Service. The lecturer says: "A Machine Gun proper should have no recoil, and should also be heavy enough to resist great vibration after being fired, so that the sighting cannot be affected by the discharge of the piece." "Such was the terror inspired by these guns (Gatlings, which pumped lead) when used for clearing the streets, that although there was an army of over 9,000 men within a short distance, they would not face the small party of 370 men," . . . at Alexandria in 1882, under Captain J. Fisher, R.N. "For brevity's sake the three classes of Machine Guns may be called: (*a*) The rifle calibre gun ·45-inch; (*b*) the torpedo defence gun 1-inch; (*c*) the shell 2-pounder gun 1·5-inch. Each of these guns should have their place in every man-of-war." *Major Rogers* said the definition of a Machine Gun is one that is loaded automatically; the high angle fire (82° at 200 yards) of which the Gatling was capable, will form a great feature in future tactics; he hoped every infantry battalion would have a Machine Gun to intensify its fire action.—F. V. L.)

Brooke, Major C. K.

"The Utilization of Rifle Fire in the Field." *J.U.S.I.*, 1883, No. cxxii., pp. 805–842, and 23 tables. (Communicated.) "Unfortunately the study of the science of 'Fire Tactics' has been so neglected in this country, and so little data has been collected, that it is necessary to appeal chiefly to the results of experiments carried out abroad for the facts on which to base a theory of how best to utilize the power of the rifle in the field. This paper will, therefore, be based upon, and make free use of, the tables, deductions, and criticisms contained in a 'Commentary on the German Musketry Instructions,' which appeared in 1882 in the *Revue Militaire de l'Etranger*. The result of the experimental firing conducted near Dungeness in 1879–80 will also be made use of as far as practicable." Field firing may be divided into (*a*) Individual, being the fire of individual men when left to their own initiative; (*b*) collective, being the fire of several men acting together under the orders of a superior. The 23 tables are most exhaustive in demonstrating results of careful and systematic fire experiments. Under "Tactical Considerations" emphasis is laid on the study of trajectories in relation to varieties of ground.—F. V. L.)

Fosbery, Lieut.-Colonel G. V., V.C.

"Magazine Rifles and Repeaters." *J.U.S.I.*, 1883, No. cxxii., pp. 777–801. 1 plate with 10 figures, Spencer-Lee Magazine Gun, Shulhof Repeater, Mannlicher Repeater, Deeley-Edge Magazine Gun, and several forms of cartridge-holders (now known as clips). Lecture June 29, 1883. Chairman, Colonel the Earl of Wemyss and March, A.D.C. The following officers took part in the discussion:—Major Fergusson, Colonel Sir L. Graham, Mr. Nordenfelt, Admiral Selwyn, Captain McEvoy, Captain Lumley, Admiral Boys, Colonel Arbuthnot, and Major-General Gordon. (*Note.*—The lecturer deals with improvement in infantry fire training, and urges great increase allowance of ammunition for annual training. Urges introduction of field shooting with a rapid, accurate aim. States that most European countries have decided for magazine rifles. *Major Fergusson:* "If we are to inculcate rapid firing as the great desideratum, I think we shall do a great deal of mischief. How are you going to carry cartridges enough?" *Colonel Fosbery:* "As to the question of the weight of ammunition to be carried by the men, a U.S. officer lately told me that during the War of Secession he commanded a brigade in which one regiment was armed with repeaters, while the others carried the single loader, that the repeaters used less ammunition than the single loaders." The men with repeaters having the confidence of an available reserve of rounds can wait till the enemy gets within 100 yards, and probably finish him with the first round, and all the rest are saved. —F. V. L.)

1884.

Beresford, Captain Lord Charles, R.N. (British.)

"Machine Guns in the Field." *J.U.S.I.*, 1885, No. cxxviii., pp. 941–963. 2 plates; the Gatling, Gardner and Nordenfelt Machine Guns; and details of the new Gatling gun. Lecture, July 4, 1884. Chairman, Major-General H. A. Smyth. The following officers took part in the discussion:—Admiral Boys, Captain Colomb, R.N., Captain Lewis, R.E., Colonel Arbuthnot, R.A., Major Rogers, Lieut.-Colonel Holley, R.A., Lord Chelmsford, Mr. Nordenfelt, Captain Johnstone, R.N., Captain Armit, and Admiral Selwyn. (*Note.*—This is a review of facts and opinions concerning Machine Guns very clearly expressed. ". . . It must be remembered that the Navy has had more actual experiences in the working of Machine Guns in the field than any other branch of Her Majesty's Service; and guns for this purpose are supplied to the Naval Service, but not to the Army." "As a seaman I will not venture to give my own opinion on a question so essentially a military one, but will give a few opinions I have received from officers attached to different arms of the Service, some of whom, while approving of the Machine Gun, are doubtful whether it could be attached to the arm they represent, but all of whom admit its utility under certain conditions." "The point I want to bring forward is the question of the Mounting. The guns are allowed to be useful under certain circumstances, and the main point is the best way of mounting them. . . ." *Lord Chelmsford:* ". . . Because whilst it has been shown that Machine Guns are capable of very great development, it is clear that that development has not at the present moment received much encouragement from military authorities."—F. V. L.)

Colt. (U.S.)

"Colt Automatic Gun, Description of." Colt's Patent Fire Arms Manufacturing Company.

Gatling. (U.S.)

"Ametralladora de Gatling." New York. 32 pp., 11 plates.

JAMES, CAPTAIN W. H., P.S.C.

"On some Changes in Tactics caused by the Increasing Power of Modern Fire." *J.U.S.I.*, 1885, No. cxxvii., pp. 925–940. 1 plate of diagrams concerning trajectories and ground. Lecture, May 23, 1884. Chairman, Lieut.-General Sir A. J. Herbert, K.C.B. The following officers took part in the discussion:—Colonel G. Moncrieff and Colonel L. Hale. (*Note.*—The lecture deals with the enlarged dangerous zone, due to the introduction of rifle fire in the period 1854–66, and the present introduction of the small bore rifle with a flatter trajectory. "Modern fire has dissolved the close formations, and the advance now consists of a number of small units, led each by an officer, guided by one general idea, but not moving as a coherent whole in the same sense as the closed masses did in former days." ". . . That on the defensive supports to a shooting line should, during the close infantry attack, at any rate, be closed up to the line they are to support; probably in deep shelter trenches within 50 or 60 yards."—F. V. L.)

NORDENFELT, THORSTEN. (Swedish.)

"The Nordenfelt Machine Guns." Griffin. 4to., 206 pp., 57 plates (*Note.*—This is a most useful book for reference.—F. V. L.)

1885

EXHIBITION. ("Engineering.") (British.)

"Artillery at the Inventions Exhibition." Hotchkiss, Gardner Nordenfelt.

ROOS, G. (Russian.)

"Emploi des Mitrailleuses et Canons à Tir Rapid." Petrograd.

VERY, LIEUT. E. W. (U.S. Navy.)

"Hotchkiss Revolving Cannon." Paris. 4to., 68 pp., 30 plates.

1886.

ARMIT, CAPTAIN, Central London Rangers. (British.)

"Machine Guns: their Use and Abuse." *J.U.S.I.*, 1886, No. cxxxiii., pp. 37–68. 1 plate, showing: Hotchkiss, 1- and 2-pounders; Hotchkiss revolving cannon mounted as a field gun; Maxim gun of an early form, and a Nordenfelt Machine Gun on Colonel Alt's magazine carriage. 1 table on concentration of fire. Chairman, Captain the Right Hon. Lord C. Beresford, C.B., R.N., M.P. The following officers took part in the discussion:—Admiral Boys, Mr. C. F. Lowe, Lieut. Tupper, R.N., Colonel Alt, Major Lockyer, R.A., Mr. T. Nordenfelt, Colonel Liddell, Captain Acland, R.A., Rear-Admiral Fremantle, C.B., Colonel Hope, V.C., and Mr. Accles. (*Note.*—This is a most sound and interesting lecture, and a very helpful discussion. The lecturer was the first officer of Volunteers to train a Machine Gun section, and at the same time he had to compile his own drill. He trained the section of two guns to act in co-operation with the firing line of the battalion, using man draft; he did not want an escort. To those who can read between the lines it is evident that Captain Armit was the first Machine Gun tactician for enclosed country. He maintained that Machine Guns should only be worked by specially trained men, and for this purpose a Machine Gun Corps, with a depot for training, should be created. There should always be 4,000 machine gunners and 250 officers at the depot, ready to man 200 guns, giving one lieutenant and twenty men per gun, and one captain to each battery of four guns.—F. V. L.)

Benson, Lieut. G. E.

"A Machine Gun Battery and Its Equipment." *J.U.S.I.*, 1886, No. cxxxiii., pp. 69–76. (Communicated.) Dated June 10, 1885, at Mhow, Central India. (*Note.*—The subject is divided into (*a*) employment of Machine Guns in a battery; (*b*) the equipment of a battery. Complete establishment of men, horses and material for both a 4- and a 6-gun battery are given. "My experience of Machine Guns is not very extensive, having been chiefly acquired in the Suakin Expedition of this year; during this time I was attached to the R.M.A. to assist them in forming a Gardner gun battery drawn by mules on field battery lines." "It came into action at Hasheen on March 20, 1885, opening fire at 200 to 300 yards."—F. V. L.)

Lloyd, Captain W. N., R.A. (British.)

"A Gardner Machine Gun Battery in Burma." *P.R.A.I.*, 6 pp.

Navy. (British.)

"Machine and Quick-firing Guns." *Naval Annual*, 1st year, pp. 370-382.

Puckerma, W., Marine Artillery Engineer. (German.)

"Quick-firing Guns." Translated by Captain E. S. May, R.A. *P.R.A.I.*

West, Major M. R., R.H.A.

"Suggestions for the Adoption and Adaptation of the Single Barrel Machine Gun for the Various Branches of Land Service." *J.U.S.I.*, 1886, No. cxxxiii., pp. 21–36. 1 plate showing one barrel Gardner Gun on artillery limber and on pack horse. Lecture November 13, 1885. Chairman, General the Right Hon. Viscount Wolseley, G.C.B., G.C.M.G., Adjutant-General of the Forces. The following officers took part in the discussion:—Lord Charles Beresford, Major-General W. Laurie, Admiral Arthur, General Smythe, R.A., Colonel Markham, R.A., followed by Mr. T. Nordenfelt explaining his gun. (*Note.*—

This contains descriptions of experiments carried out by the lecturer; firing, from a gun limber on the move and at the halt; carrying a Machine Gun and tripod on a pack horse for cavalry or infantry. Quotations are made from reports of lectures and Continental experiments; also from senior officers of experience. *Chairman* : " It is a subject I think especially for discussion, because the English Army has now most certainly arrived at the conclusion that we must have Machine Guns, and I am very glad to say the authorities have at last decided upon their being introduced into the Army." " . . . But there certainly is a very common impression in the minds of a great number, that opposition from Woolwich has prevented our having . . . Machine Guns for many years past." " But I feel convinced the fire of this small arm, an infantry arm—it is not an artillery arm—I think the fire of this small arm, firing from a fixed carriage at ascertained ranges of 2,000 up to 3,000 yards and beyond, will be most effective."—F. V. L.)

1887.

ANDERSON, MAJOR A. D., R.H.A. (British.)

" Suggestions as to the Use of Machine Guns in the Field, in Combination with Infantry." *J.U.S.I.*, 1887, No. 138, pp. 45–56. 1 sketch map. Lecture on January 21, 1887. (Read, in the absence of Major Anderson, by Major Hime, R.A.) Chairman, Lieut.-General Sir E. B. Hamley, K.C.B., K.C.M.G., R.A., M.P. The following officers took part in the discussion:—General Lord Chelmsford, Major H. L. Hime, Colonel Eardley Wilmot, Sir W. Olphert, V.C., K.C.B., Colonel Baylis (of Volunteers), Colonel Richardson, R.A., Colonel Markham, R.A. (*Note.*—The lecturer comes to the conclusion that the true tactics of the Machine Gun with infantry are (*a*) the closest connection between the guns and battalions; (*b*) the guns to aid and support the firing line; (*c*) the guns to be pushed forward as freely and boldly as a section of a company. In regard to the mounting and transport, he considers the above tactics can only be carried out, with tripod mounting and pack transport, one mule for gun, mounting and spare parts, and one for ammunition. There should be eight guns per battalion, one for each company (under the captain), of which four men are to be trained in its use. At any moment, however, the battalion, brigade or divisional commander would be in a position to order the whole, or a portion of them, to mass on some suitable spot in support of an advance, an attack, or threatened point. *Chairman* : " . . . We must remember that the Machine Gun, after all, is only taking the place of the infantry arm—that is, merely so many rifle barrels discharging rifle ammunition, and, therefore, the control of them may properly, I think, be entrusted to an infantry commander."—F. V. L.)

Benson, Lieut. G. E., R.A. (British.)

"Machine Guns: their Tactics and Equipment." *J.U.S.I.*, 1888, No. 142, pp. 937–972. 1 plate, showing trail mounting for artillery, pair horse cart mounting for cavalry, and two wheel one-mule draft mounting for infantry; all these were for Nordenfelt Machine Gun. Lecture, October 18, 1887. Chairman, Major-General W. H. Goodenough, C.B., R.A. The following officers took part in the discussion:—Colonel the Hon. R. A. T. Talbot, C.B., Major W. W. M. Smith, R.A., Captain Stone, R.A., Lord Chelmsford, Mr. Maxim, Lieut.-Colonel Alt, Mr. C. F. Lowe, Colonel T. Coke, Captain W. H. James, Mr. Nordenfelt, Major-General Arbuthnot, Lieut.-Colonel J. P. Brabazon, Captain R. S. Baden-Powell, Captain Palliser, and Lord C. Beresford. (*Note.*—This is a most important lecture and discussion. In comparison with former papers, speakers kept well to the point. There is a good résumé. Lieut. Benson's proposal is: Three troops of four guns to each division of infantry, the guns to be on trail mountings with limbers and four horses each, four ammunition wagons or carts per troop. ". . . What we want is to try their combined effect under service conditions and at targets representing such objects as would be fired at on service." *Major Smith :* "In my opinion, the Machine Gun ought to act with the brigade of infantry." "I estimate for a 9-gun battery with seventy-five to eighty horses and 130 men, mounted on trail and limber, but only two draft horses." He urges a school of the arm to be formed at Aldershot, and that after enough field experiments had been carried out it would become a nucleus for the future Corps of Machine Guns. *Captain Stone* supported the idea of training and working Machine Guns as a separate arm. *Colonel Alt* claimed better co-operation with infantry for his man draft mounting than for horse draft. *General Arbuthnot* said all the Machine Guns of the Service had passed through his hands. He thought there should be one equipment for cavalry and one for infantry; further, the infantry should have a tripod pack equipment for mountainous countries and a wheeled landing carriage for open and cultivated country. He laid down no Machine Gun should exceed 100 lbs. in weight, so that it could be used with a light carriage. (The Vickers' Gun, with water in casing, weighs 38½ lbs.) *Lord C. Beresford* laid great emphasis on the correct term being "Machine Rifles." Here is a most prophetic sentence: "It would be a very serious thing if the German, or any other, Army were to take up the Machine Gun question, and we, with all our practical experience, having found it so useful on so many occasions, were not to take it up and thrash it out as has been proposed." *Lieut. Benson*, in conclusion, considered Machine Guns must be formed into troops for administrative (and training) purposes. It is easy enough to get forces to act separately which have been trained to act together, but it is a very different matter to get those to act together which have only been taught to act separately.—F. V. L.)

JAMES, CAPTAIN W. H.

" Magazine and Repeating Rifles." *J.U.S.I.*, 1887, No. 138, pp. 135–161. 3 large plates of rifles and trajectory curves; Mauser, Kropatscheck, Mannlicher, Jarman, Vetterli, Owen-Jones and Schulhof. Lecture, February 25, 1887. Chairman, Major-General E. H. Clive, Commandant Staff College. The following officers took part in the discussion:—Major-General Bray, C.B., Mr. C. F. Lowe, Major Mackinlay, Colonel Arbuthnot, Colonel L. Hale, Colonel Fraser, Admiral Selwyn, Admiral Boys, Captain P. H. Colomb, R.N., Lieut. Tupper, R.N., Major-General Dunn, and Major Leslie. (*Note.*—The lecturer deals with (*a*) the tactical, and (*b*) the technical considerations. In the former he advocates sudden showers of bullets in volleys, which can only be obtained by the use of a magazine; this rate of fire only to be used at short ranges, though it could be used up to 1,500 yards. " Such weapons require careful training both of the officers and men, frequent practice in their use, and careful working out of the problems they give rise to." Under the heading of (*b*) the technical considerations, " It will be seen, therefore, that every nation has either definitely adopted, or is experimenting with a view to deciding on, some form of magazine." The British Army authorities are experimenting with a view to deciding on some form of magazine rifle. The lecturer complains that the British System of Musketry training is not equal to that of France, Germany or Russia.—F. V. L.

KOERNER, A. C. (U.S.)

" The Hotchkiss Single Barrel Rapid-firing Gun." Paris. 37 pp., 10 plates.

TCHEBICHEF, GENERAL. (Russian.)

In an essay on the " Mitrailleuses," published by the *Ruszki Invalid* in 1887, General Tchebichef, Professor at the Artillery School, Michel, cites two occasions in the war, 1877-78, in which these weapons (in this case they were Gatlings) played an important part, and he concludes that an army should always be provided with them. The first occasion was at the siege of Nicopolis by the Russians, who opened fire at 900 metres against the Turks with great effect. The second occasion was at the siege of Plevna, at 1,000 metres, also with quick effect.

1888.

HAMILTON, W. R. (U.S.A.)

" American Machine Cannon and Dynamite Guns." New York. (*Note.*—A popular magazine article.—F. V. L.)

MAYNE, CAPTAIN C. B., R.E.

"Infantry Fire Tactics." Gale and Polden, Chatham. Second Edition, 1888, dated at the Royal Military College, Kingston, Ontario, Canada. 8vo, pp. xii and 526. 4 tables and 1 trajectory plate. A bibliography, but no index. (*Note.*—The first edition was published in 1884. This work easily stands out head and shoulders above all contemporary works on the subject, and had a great influence on the Official British Manuals and Drill Books for Infantry. It has long done its work; but it was so highly thought of by Lord Wolseley that he presented a copy of it in 1905 to the library of the Royal United Service Institution.—F. V. L.)

MECHAM, MAJOR J. R. (British.)

"Trials between Maxim and New Service Rifle." *P.R.A.I.*, 3 pp.

ROWAN, LIEUT. T. T., R.A. (British.)

"Small Arms *v.* Machine Guns." *P.R.A.I.*, 1 p. (At Singapore, October, 1888.)

STONE, CAPTAIN F. G., R.A. (British.)

"The Maxim Machine Gun." *A.M.S.*, 1888. Lecture No. 7. Reprinted in *P.R.A.I.*, vol. xvi., pp. 615–621. (Owing to the *A.M.S.* lecture being out of print, it is not possible to give names of Chairman and officers who took part in the discussion.) (*Note.*—Throughout this lecture, the Machine Gun fire is still considered as shrapnel fire in effect, rather than as the Infantry Bullet Sheaf. The gun should be mounted in various ways to suit different conditions (of ground), on tripod or travelling carriage. Lieut. Benson is agreed with, that Machine Gun troops of four guns be formed, and that they should be attached on the same lines which are used in the distribution of batteries of horse and field artillery. Proposed sixteen guns per infantry division, four guns per cavalry brigade, four per cavalry division, and twenty-four per army corps. One pattern carriage for all purposes, limber and trail with shield with four horses, 7,992 rounds (·45-inch) on trail and limber, one non-commissioned officer, two gunners and two drivers. The troop to contain captain, lieutenant, serjeant-major, Q.M. Serjeant, and farrier, total with non-commissioned officers, gunners and drivers, of twenty-five officers and men—F. V. L.)

1889.

DITTRICH, CAPTAIN A. (German.)

"Machine Guns and their Employment." *P.R.A.I.*, 16 pp. Translated from the German by Captain E. S. May, R.A.

ROGERS, LIEUT.-COLONEL E. (British.)

"Machine Rifle Batteries for Volunteers." Manchester, J. Heywood. 8vo., 60 pp., 13 plates.

STONE, CAPTAIN F. G., R.A. (British.)

"A Description of Quick-firing Guns." *P.R.A.I.*, 24 pp. Definitions.

1891.

CHELSEA.

"Catalogue of the Royal Naval Exhibition held at Chelsea in 1891." P. 453, "Gatling, Gardner, and Nordenfelt Machine Guns"; p. 468, "The Maxim-Nordenfelt Gun and Ammunition Company, Ltd."

HOTCHKISS. (British.)

"Hotchkiss Gun Works." *Engineering*, January 23, 1891. 3 pp., plates.

1892.

LANGLOIS, COLONEL. (French.)

"L'Artillerie de Campagne en Liaison avec les Autre Armes." 2 vols. Baudoin.

MORGAN, MAJOR F. C., R.A. (British.)

"Handbook of Artillery Material." Clowes. pp. 120-130.

"Machine Guns." Gatling, Gardner, Nordenfelt, and Maxim.

1894.

THOMSON, CAPTAIN A. W., R.H.G. (British.)

"Machine Guns with Cavalry." *J.U.S.I.*, vol. 38, 1894, pp. 617–639. 1 plate showing Non-Recoil Galloping Carriage with three horse quick-firing Nordenfelt. One map of the Idstone Manœuvres of 1893. Lecture, February 23, 1894. Chairman, Lieut.-General J. K. Fraser. The following officers took part in the discussion:—Major Beauchamp, Colonel J. B. B. Dickson, Captain East, Major G. Stone and Colonel L. Hale. (*Note.*—The lecturer had command of a Maxim on a cavalry mounting for six days at the Idstone manœuvres in 1893, and he relates his experiences of each day, and suggests possible improvements in his tactics. Under Material, he deals with (*a*) transport; (*b*) ammunition; (*c*) range; (*d*) sights; (*e*) shield. Under (*a*) he urges an experiment with a four-wheeled buggy with gun mounted close to rear axle. At the same time a pack saddle and tripod should be carried for rough country. The Cape Mounted Rifles have adopted this suggestion, and a pack saddle is to accompany every Maxim. In addition each gun should have an extra horse and saddle for ammunition. Mention is made of the twenty-four Maxims worked in the Swiss cavalry. An extract of a letter from Major Baden-Powell was read by the Chairman, in which he points out that Machine Guns ought to be as low down as the axle; that a tripod should always be carried, and that the gun should be easily detachable to be placed on it.—F. V. L.)

1895.

LAKE, V. R. (British.)

"Patents for Inventions relating to Machine Guns." 106 pp. Published by Haseltine, Lake and Co., London.

MAXIM, H. S. (British.)

"News Cuttings and Illustrations." London.

1896.

MAXIM, H. S. (British.)

"Letter to Messrs. Ludwig Loewe and Co." 24 pp. *The Bedford Press.*

WILLE, MAJOR-GENERAL R. (German.)

"Textbook on Machine Gun Construction." Published by Seidel. First edition.

1897.

STIPSIEZ. (Austrian.)

"Employment of Machine Guns by Cavalry." Vienna.

1898.

"AJAX" (Captain H. P. de la Bere). (British.)

"Machine Guns, their Use and Abuse." *U.S.M.*, August, 8 pp.

LOCKYER, LIEUT.-COLONEL W. N., R.A. (British.)

"Personal Reminiscences of the Evolution of Small Arms and Machine Guns from the Year 1863 up to the present Day." *J.U.S.I.*, 1898, No. 248, pp. 1121–1151. No plates. Lecture, June 29, 1898. Chairman, Rear-Admiral the Right Hon. Lord Charles Beresford, C.B., M.P. The following officers took part in the discussion:—Major J. Formby, Lieut.-Colonel E. Gunther, Captain W. James and Major-General H. T. Arbuthnot. (*Note.*—The lecture is divided into (1) Rifles; (2) pistols, swords, bayonets and Machine Guns; (3) inspection. Machine Guns were introduced into the Army as follows:—Gatling, in 1875, weighing 416 lbs.; Nordenfelt, in 1878, of 3, 5, and 10 barrels; Gardner, in 1882, of 1, 2, and 5 barrels; Maxim, in 1887, weighing 60 lbs. (Vickers' Machine Gun weighs 38½ lbs.). The position of the lecturer as Chief Inspector of Small Arms gives great value to his opinions on Machine Guns. He thinks the Maxim is a good Machine Gun, but it can hardly be called an ideal automatic gun, and he believes that the evolution of this arm is still in progress. He considers that the qualities to be desired in a Machine Gun are in the following order of merit:—(1) Certainty of action; (2) simplicity, that is, fewest parts; (3) endurance; (4) accuracy—extreme accuracy is not absolute y necessary for a Machine Gun in the field; you do not want to put twenty or thirty bullets through the same man, and an automatic gun with a lateral spread would be rather a good thing. Accuracy depends upon the rifling and steadiness of the mounting.—F. V. L.)

1899.

ALGER, PROFESSOR P. R., AND E. N. C. TWINY. (U.S. Navy.)
"The Colt Automatic Gun, Calibre 6 mm., Mark I." London. Pamphlet.

GALE. (British.)
"The Maxim Gun: Questions and Answers." First edition by Gale and Polden.

MISSY, COMMANDANT H. DE. (French.)
"The American Machine Guns at Santiago." 20 pp.

PARKER, LIEUT. J. H., 13th U.S. Infantry. (U.S.)
"Tactical Organization and Uses of Machine Guns in the Field." Kansas City.

PARKER, LIEUT.
"Machine Guns in the Spanish-American War." *U.S.M.*, March.

PARKER, LIEUT.
"The Gatling Guns at Santiago." Introduction by Colonel T. Roosevelt, 1st U.S. Volunteer Cavalry. 300 pp., 37 photographs, and 2 plans. Hudson Kimberley Publishing Company. (*Note.*—A very interesting pioneering Machine Gun story, full of great difficulties overcome.—F. V. L.)

1900.

BERE, CAPTAIN H. P. DE LA, Royal Scots Fusiliers. (British.)
"Machine Guns in Tirah." *U.S.M.*, November, 8 pp.

BOURDON, J. (French.)
"Notes on Machine Guns." *R.d'A.*, July.

LUZEUX, GENERAL A. (French.)
"Machine Guns in Modern Warfare." Lavauzelle, 8vo., 2f.

1901.

GUNTHER, DR. R. (German.)
"La Mitrailleuse Bergman." Rastatt.

NORDENFELT. (British.)
"Nordenfelt's Quick-Firing Field Artillery." Société Cockerill, Seraing. *Engineering*, March 1, 1901, pp. 259-262.

1902.

LAVAUZELLE. (German.)

"Projet de Reglement de Manœuvres pour les Groupes de Mitrailleuses." 124 pp. Translation published by H. C. Lavauzelle.

ROUQUEROL, COMMANDANT G. (French.)

"Organization of Quick-Firing Field Artillery." Berger-Levrault, vol. i., pp. 439-456. Contains a chapter on Machine Guns. English translation published by Hugh Rees, Ltd.

SARASIN, CAPTAIN C. (Swiss.)

"Machine Guns in the German Army." *R.M.S.*, November, p. 942.

SCHLEYER, COLONEL. (Swiss.)

"The Progress of Machine Guns." A lecture printed in *R.M.S.*, March, p. 262.

1903.

BRAUN, CAPTAIN D. VON. (German.)

"La Mitrailleuse Maxim et Son Emploi." Second edition, 118 pp. Translated from the German. Eisenschmidt.

HOTCHKISS. (British.)

Engineering, February 20, pp. 241-243, 277-280, 333-337, 374-377, 437-438. The Hotchkiss Ordnance Company, Ltd., St. Denis.

MAYNE, LIEUT.-COLONEL C. B., R.E.

"The Infantry Weapon and its Use in War." Smith, Elder and Co., 1903, pp. xii and 341. Index. (*Note.*—This is the best textbook on Infantry Fire Tactics, and has not been superseded. Some extracts are given. "This book has been written in response to the numerous requests that have been received at various times and from various quarters for a revision of my book on 'Infantry Fire Tactics.' But while correcting the proof-sheets, it has occurred to me that I have not laid sufficient stress on the necessity for a close and continuous co-operation of the artillery and infantry arms in battle, since, especially nowadays, each arm requires the assistance of the other arm to enable it to do its work of giving assistance to this other arm. My work on 'Infantry Fire Tactics' was first published in 1884. . . . Its primary object was to plead for the introduction into our Regulations of many important usages and methods of firing that were necessary for insuring the proper use of that (Martini) particular rifle in war."—F. V. L.)

1904.

BIRD, LIEUT.-COLONEL W. D., D.S.O. (British.)

"Rifle Calibre Machine Gun Tactics." *A.M.S.*, No. lxxx., 15 pp. Lecture, November 8, 1904. Chairman, Major-General A. H. Paget, C.V.O., C.B. The following officers took part in the discussion:—

Brigadier-General E. A. H. Alderson, C.B., A.D.C., Brigadier-General H. E. Belfield, C.B., D.S.O., and Major-General F. Eustace, C.B. (*Note*.—The lecturer favours the tripod mounting and carried on pack-horse, together with the brigading of the guns. He considers the Machine Gun is merely a machine rifle, and it condenses in one arm, with the fewest possible accessories, the power of a certain number (50–60) of rifles. It is not possible to institute a satisfactory comparison even with the rifle on account of the difficulty of computing the moral factors which enter so largely into the case. Its potentialities cannot be fully utilized until its power of rapid and accurate fire is fully realized. The characteristics of the Machine Gun point in favour of collective action, which facilitates control, direction and concentration of fire. Whether training should be collective or isolated, is a question beyond the scope of this lecture. In order to obtain the best results, the immediate commander should possess a high standard of technical knowledge, a quick eye for ground, the power of rapid grasp of the situation and bold action. *General Alderson:* He considers that a text-book on Machine Guns, and especially their tactics, was very much wanted. He advocates the tripod and pack-saddle. He considers that the good tactical handling—and this is quite as, if not more, important than good shooting of Machine Guns—is only learnt by experience. He advocates the men should shoot more often and be allowed much more ammunition. He considers we have enough experience to work upon ourselves without going to the Germans.—F. V. L.)

KYNOCH. (British.)

"History of the Cartridge." *Kynoch Journal*, 1904, pp. 29, 92, 155 182. (*Note*.—Very good.—F. V. L.)

MLENEK, CAPTAIN. (French.)

"Notes on the Machine Gun." Lavauzelle. 140 pp., 46 illustrations.

ROGERS, LIEUT.-COLONEL E.

"Machine Guns Up-to-Date." *J.U.S.I.*, 1904, No. 319, pp. 1032–1049. Lecture, April 29, 1904. Chairman, Colonel G. W. A. Fitz-George. The following took part in the discussion:—T. Miller Maguire, M.A., Sir H. Maxim, and Lieut. J. W. Dunne. (*Note*.—The first few pages contain a résumé from 1872 up to Lieut. Parker at Santiago in 1899. The main part deals with the use of the Maxim, Colt and Pom-pom in the South African War, 1899–1902. There are many valuable notes and extracts describing the equipment, organization and tactics used on both sides in the war. Sir H. Maxim sheds some more light on the creation of the Pom-pom.—F. V. L.)

VUILLEUMIER, CAPTAIN. (Swiss.)

"Le combat entre l'infanterie et les mitrailleuses." From *R.M.S.* September and October. (*Note*.—This is a long and valuable paper. —F. V. L.)

1905.

BRAUN, CAPTAIN A. D. VON. (German.)
"Das Maxim-Maschinengewehr und seine Verwendung." V. von R. Eisenschmidt, Berlin, 1905. 144 pp., 19 folding plates at end of book, many illustrations and plans in the text. A translation of the Preface to Third Edition reads as follows: "As the Second Edition of the 'Maxim Machine Gun and its Use,' by Captain A. D. von Braun, is exhausted, a new edition has become necessary. This, at the request of the German Arms and Munitions Company, has been revised, and differs in form and contents from the Second Edition, from which only some extracts have been taken unaltered. Dated May, 1905. . . . A number of both drawings and photographs have been taken from this book for reproduction in 'The Book of the Machine Gun.'"

EXLER. (Austrian.)
"Musketry Training for Small Arms and Machine Guns." Seidel.

GENNARO, LIEUT. R. DE. (Italian.)
"A Study on Machine Guns."

KORZEN AND KUHN. (Austrian.)
"Study of Weapons." Vol. VIII. Machine Guns. Seidel.

LAYRIZ, LIEUT.-COLONEL VON. (German.)
"The Advantages which the Machine Gun Assures to the Three Arms." *K.T.Z.*

ROHNE, LIEUT.-GENERAL. (German.)
"Field Firing of a Detachment of Infantry and Firing of a Machine Gun."

WILLE, VON R., MAJOR-GENERAL. (German.)
"Waffenlehre" (meaning "Weapons"). 3 vols. Berlin, Eisenschmidt. Vol. i., pp. 152-178, deals with Machine Guns. There is a large folding plate with many photographs of Machine Guns. The information given is chiefly technical. On p. 287 in vol. iii. is a very good bibliography, which is followed by the index for all three volumes. On p. 299, under Machine Guns, there are 47 bibliographical entries. (*Note.*—This work is the one studied largely by German officers for their professional training. It compares all kinds of weapons in use by the principal armies of the world—guns, rifles, pistols, etc. There is a real want of such a work in English.—F. V. L.)

1906.

ANISIMOV, LIEUT.-COLONEL. (Russian.)
"Les Mitrailleuses." (*Note.*—Very good.—F. V. L.)

GUERIN, COMMANDANT. (French.)
"Emploi des Mitrailleuses dans l'Armament des Troupes Metropolitaines et Coloniales." Lavauzelle. 68 pp.

MUNIER, COMMANDANT. (French.)
"A Study of the Machine Gun." From *J.D.A.M.*, November and December.

NOEL, LIEUT D'E. M. (Belgian.)
"Machine Guns." *R.M.B.*, May to December.

P. DE L. (French.)
"Machine Gunners and Machine Guns." *La Patrie*, March 15, 1906. (*Note.*—Good for mountings.—F. V. L.)

STEIGER. (Swiss.)
"Machine Guns in the Infantry and Cavalry." Zurich.

1907.

ALGRAIN, LIEUT.-COLONEL A. (Belgian.)
"Reinforced Concrete in the Construction of Fortifications." 20 pp

ALIPRANDI, LIEUT. A. H. (Austrian.)
"Cavalry Machine Gun Sections." Seidel.

APPLIN, CAPTAIN R. V. K. (British.)
"Machine Guns with Cavalry." *C.J.*, July.

BALCK. (German.)
"Machine Guns and their Employment." *J.D.A.M.*, April.

BINDER, LIEUT. F. (Austrian.)
"The Machine Gun." Seidel.

BUCALO, CAPTAIN G. (Italian.)
"Notes on the Tactical Employment of Machine Guns." *R.M.I.*, May.

CAVALRY. (Austrian.)
"The Equipment of Cavalry with Machine Guns." *K.M.*, July.

CZERLIEN, MAJOR-GENERAL V. M. (Austrian.)
"The Employment of Machine Guns in Conjunction with Cavalry." *K.M.*, August and September.

GAUTREAU, CAPTAIN. (French.)
"A Lecture on Machine Guns."

HARTMAN, E. (German.)
"Machine Guns and their Use in War." Mittler. (*Note.*—This is to the glory of the Maxim.—F. V. L.)

HEYDENREICH, LIEUT.-COLONEL. (German.)
"The Principal Machine Guns, according to Trials in Different Countries." *A.M.B.*, March.

HUMBERT, C. (French.)
"The Question Humbert." Paris. (Hotchkiss *v.* Puteaux Guns.)

IMMANUEL, MAJOR. (Austrian.)
"The Organization of Cavalry Machine Gun Sections." *K.M.*, August and September.

KORZEN, A. (Austrian.)
"The New Schwarzlose Machine Gun, Model 1907." *O.M.Z.*, March.

NOBELL, VON. (German.)
"Annual Report on Armies and Military Matters." 495 pp.

MCMAHON, LIEUT.-COLONEL N. R., D.S.O. (British.)
"Fire Fighting." *A.M.S.*, December 18. London, Rees. Major-General J. M. Grierson, C.V.O., C.B., C.M.G., in the chair. The following officers took part in the discussion: Brigadier-General C. J. Mackenzie, C.B., Brigadier-General F. Lloyd, C.B., D.S.O., Colonel G. G. A. Egerton, C.B., Colonel D. Henderson, D.S.O., and Captain A. Bryant. At this time Lieut.-Colonel McMahon was the Chief Instructor at Hythe, and Colonel Egerton was the Commandant of the School of Musketry. This is a most important historical study which shows an active period in the formation of British Machine Gun Tactics. He regards the Machine Gun as being an automatic rifle, and advocates converging fire whether from the front or a flank. Intervals between large bodies of troops in the attack are good positions for Machine Guns. Successful combination might take the form of an established system of rapid ground reconnaissance, a careful allotment of duties with fire, under the heads of (1) Organization; (2) Direction; (3) Control; (4) Discipline: a definite system of selecting targets based on general principles for distribution of covering fire and concentration of decisive fire. The means are specified in the books, but great things are necessarily mixed up with small, and those which have general application are indistinguishable from those of limited bearing; fire effect cannot be studied at manœuvres; it can best be dealt with theoretically at musketry conferences, practically on rifle ranges. "If attention is confined to a certain area of ground, the quick application of fire and judging distance is facilitated." "The case in which the volume of fire is restricted by frontage is one of many in which Machine Guns will be invaluable." "If Machine Guns are massed with us there is no chain of command." Our own great difficulty is want of GROUND for the study of fire direction and effect; the areas should be wide enough to force officers to use field glasses for fire direction and effect all the time. "Our traditions of rapid and accurate fire date back nearly 600 years." *Chairman:* ". . . The necessity is for our not repressing, but rather directing, the initiative of our company officers. What is wanted is to direct that initiative into suitable channels and to give them general principles to work on . . . for the common good."—F. V. L.

SCHWARZLOSE. (French.)
"The Schwarzlose Machine Gun." *Revue d'Infanterie*, July 15, pp. 14-20.

TAKENOUCHI, CAPTAIN. (Japanese.)
" On the Tactical Employment of Machine Guns with Infantry." *J.U.S.I.*

VIKTORIN, CAPTAIN H. (Austrian.)
" Pack or Wheel Transport for Machine Guns." *K.M.*, August and September.

WILSON, MAJOR C. H. (British.)
" Machine Guns as Skirmishers." *U.S.M.*, August.

1908.

BALCK. (German.)
" Tactics." Eisenschmidt, Berlin. Fourth edition, vol. i., 420 pp. English translation published by Hugh Rees, Ltd.

BECKMAN, CAPTAIN. (German.)
" The Machine Gun Question." *J.D.A.M.*, April.

BOULLE, LIEUT. (French.)
" Machine Guns in France and Foreign Countries." Lavauzelle. 60 pp.
" Machine Guns in Foreign Armies." Berger-Levrault. 77 pp.

BYERN, MAJOR. (Austrian.)
" Appendix to the Training Regulations for the Imperial and Royal Cavalry Machine Gun Detachments." Official, Vienna.

BYERN, MAJOR. (Austrian.)
" Appendix to the Training Regulations for the Imperial and Royal Infantry Machine Gun Detachments." Official, Vienna.

BYUNTING, COLONEL.
" Cavalry and Machine Guns." An abbreviated translation of an article by Colonel von Byunting in the Russian Cavalry Journal, Nos. 20 and 21 of 1908. See *R.P.M.I.*, No. 10, p. 26, for précis.

CULMAN, CAPTAIN. (French.)
" Choses d'Allemagne"—Nouvel Armament, etc. " Thèmes Tac tiques"—Emploi des Mitrailleuses, etc.

DASHKOV, COUNT. (Russian.)
" Instructions for the Tactical Training of Troops." Dated July 23 1908. By Count Vorontsov-Dashkov, General C. in C., the Caucasian Military District. Translated from the *Voenni Sbornik* of November 1908. *R.P.M.I.*, No. 10, pp. 40-48.

FIRE PRACTICE. (Austrian.)
Practice with Machine Guns." *O.M.Z.*, August, 1908.

FLECK, CAPTAIN A. VON. (German.)
"Maschinengewehre ihre Technik und Taktik." E. S. Mittler und Sohn, Berlin. 165 pp., 50 plates, and many diagrams, illustrations and plans in the text. For such a small book this is very comprehensive. It has a Bibliography of 88 entries, mostly German, French, and Austrian works and papers. Captain Fleck deals by means of diagrams and text with the following systems: Maxim, Hotchkiss, Schwarzlose, Madsen, Bergman, Colt, Skoda, Fitzgerald, and discusses the future type for aircraft. Pages 100 to 164 deal with the Tactics.

GODTS, MAJOR J. (Belgian.)
"Machine Guns. Their Employment, More Especially from a Tactical Point of View, in Co-operation with the Other Arms." Laeken, 155 pp., 8vo. Compiled from lectures.

ITALY. (Italian.)
The *Rivista di Artigliera e Genio* for the months of January, February, March, May, June, August, and September, has many articles on Machine Gun subjects.

JAPANESE REGULATIONS.
"Japanese Regulations for Machine Gun Sections." *M.W.B.*, May 12, 1908.

JAPANESE REGULATIONS.
"Japanese Regulations for Machine Guns for Infantry and Cavalry." *Revue d'Infanterie*, March 15, 1908. See Lavau, vol. ii., pp. 587-601

JAPANESE TACTICS.
"Japanese Opinions on the Employment of Machine Guns" *B.P.B.M.*, May 15, 1908.

KORZEN-KUHN. (Austrian.)
"Machine Guns." In vol. viii. of *Waffenlehre*. Seidel.

LAVAU, CAPTAIN J. C. (French.)
"Machine Guns with Cavalry." Vol. i., 657 pp., 100 engravings in the text, 8vo. Angers. This is the most useful book of reference that there is for Machine Guns.

"M." LIEUT. (French.)
"Machine Guns in Foreign Armies. Tactics and Organization." Preface by Commandant Niessel. Berger-Levrault.

MAIRETET, CAPTAIN. (French.)
"The Infantry Machine Gun: Its History and Tactical Employment." Lavau, vol. ii., pp. 504-518. *J.S.M.*, September 15, October 1 and 15, 1908.

MARSELLA, CAPTAIN. (Spanish.)
"Estudios acerca de Ametralladoras." Madrid.

MATÉRIEL. (French.)

"Engins Moderns—Mitrailleuses." Fr. Mil. Biblio., July 7, 1908; 172 figures grouped in 26 tables.

MAXIM. (Belgian.)

"The Maxim Gun, Light Pattern." Brussels, *L.B.M.*, July 19, 1908.

MERKATZ, LIEUT. F. VON. (German.)

"Instructional Book for Machine Gun Batteries." Eisenschmidt Fourth edition, 232 pp. It is a popular book for recruits and contains a great deal of useful information. It is in five parts.

NICOLAS. (Russian.)

"Conférences sur la Guerre Russo-Japonaise, faites à l'Académie." Traduites du Russe. Seven volumes. Lavauzelle. There are many passages relative to Machine Guns.

NIESSEL, COMMANDANT. (French.)

"Tendances Actuelles de la Cavalerie Allemande." Second edition. Lavauzelle. There is a chapter on Machine Guns.

PAINVIN, MAJOR. (French.)

"The New Japanese Regulations for the Employment of Machine Guns with Infantry and Cavalry." Paris, *L.R.I.*, March 15, 1908.

PAINVIN, MAJOR. (French.)

"The Organization of Machine Guns in Austria." Paris, *L.R.I.*, April 15, 1908.

PARKER, LIEUT. J. H. (U.S.)

"Progress in Machine Gun Development." *U.S.I.*, vol. v.

PELLOUX, GENERAL. (French.)

"German Field Service, Regulations of." March 22, 1908. In French. Berger-Levrault. Lavau, vol. ii., pp. 180-182.

QUESTION, THE. (Belgian.)

"The Question of Machine Guns." Brussels, *L.B.M.*, May 24, 1908.

RUDEANU, CAPTAIN V. (Rumanian.)

"Mitraliere si Pusci Mitraliere." Bukharest, A. Baer.

SAINT-ÉTIENNE. (French.)

"The New Machine Gun." *L.F.M.*, February 29, 1908.

SANDER, CAPTAIN H. (German.)

"The Employment of Machine Guns by the Japanese." *M.W.B.*, supplement 109, April, 1908, *I.R.*

SCHELLMAN, CAPTAIN V. (Austrian.)

"Machine Gun Systems and the Equipments of Machine Guns" *O.M.Z.*, June, 1908.

SIEGRINGEN, CAPTAIN K. VON. (Austrian.)

"The Tactical Employment of Machine Guns in Attack and Defence." *M.A.G.*, April.

SJÖRGREN.

"The Normal Automatic Gun, Sjörgren System." *Engineering*, July 24, 1908.

SURIN, COLONEL. (Russian.)

"Machine Gun Tactics in the Russo-Japanese War." *Russian Military Scientific Society*, No. 3 of 1908. See *R.P.M.I.*, No. 11, pp. 3-8. (*Note.*—Very good.—F. V. L.)

TACTICS. (Austrian.)

"Tactical Training of Machine Gun Detachments." *O.M.Z.*, September.

TACTICS. (Belgian.)

"The Use of Machine Guns in Action." *R.M.B.* See the *P.R.A.I.* of August, 1908.

TACTICS. (French.)

"Machine Guns." (A series of articles in the paper *L'Armée Moderne*, signed by a Colonel of Artillery, starting May 6. It is of great interest to all Arms. See Lavau, vol. ii., pp. 437-458.)

VILLAMIL, CAPTAIN. (Spanish.)

"Ametralladora Hotchkiss de Campaña en Servicio." Madrid.

YEAR-BOOK. (German.)

"Jahrbücher für die Deutsche Armee und Marine." April. (*Note.*—This contains a very important study on Machine Guns.—F. V. L.)

1909.

ARMOURED CARS. (French.)

"Armoured Machine Guns for the Pursuit of Dirigible Balloons." *Revue du Genie Militaire*, Paris, March.

AUSTRIAN.

"Machine Guns." *M.A.G.*, February.

BOERNER, LIEUT. (French.)

"The Combat and the Tactical Employment of the Infantry Machine Gun"; "Rifles and Machine Guns." A series of very good articles in the *Armée Moderne*, from July to September.

BOUCOMONT, LIEUT. (French.)

See Villemont, Captain, 1909.

BUTTIN, LIEUT. (French.)

"The Employment of Machine Guns of Infantry on the Offensive." Chapelot, 122 pp.

CAVALRY. (Austrian.)

"Considerations as to the Augmented Value of Cavalry through the Introduction of a Better Firearm and of Machine Gun Sections." *K.M.*, February.

CAVALRY. (Austrian.)

"Machine Gun Squadrons." *K.M.*, September.

C. T. (Belgian.)

"Automatic Rifles." *L.B.M.*, February 14.

FEDOROV, S. (Russian.)

"Machine Gun Questions." *V.S.*, July and August.

FEDOROV, S. (Russian.)

"Machine Gun Questions continued: The Training of the Personnel. Comparison of the Various Systems of the Machine Guns in Use by the Armies of the Present Day." *V.S.*, No. 8, p. 83.

GERMANY. (French.)

"Machine Guns in Germany." *L.R.I.*, July 15.

GOBILLOT, LIEUT. (French.)

"The Employment of the Machine Gun in the Offensive Combat." Berger-Levrault. 40 pp.

HOTCHKISS. (German.)

"The Hotchkiss Machine Gun." Berlin, *M.W.B.*, December 21, p. 3641.

KEEN, CAPTAIN F. S. (British.)

"Machine Guns." Simla, *P.U.S.I.*, October.

KORZEN, A. (Austrian.)

"The New Machine Gun M. 9, System, Schwarzlose." W. Hamburger, Vienna; *R.M.S.*, April.

KRETZSCHMAR, LIEUT. (German.)

"Practical Hints for the Training of Machine Gun Companies from the Musketry Point of View." Stalling, *R.P.M.I.* contain a very good précis of the above.

LAVAU, CAPTAIN J. C. (French.)

"Machine Guns in Co-operation with Cavalry." Angers, published by Siraudeau; vol. ii., 916 pp., about 100 engravings in the text. (*Note.*—The study for this book was started in 1902; the first volume was printed in June, 1908, and the second in December, 1909. This book is a veritable compendium on Machine Guns in every way; unfortunately it has no index, and therefore much patience is required to get all there is to be got out of it. So far it has not been translated into English.—F. V. L.)

LONGSTAFF, CAPTAIN F. V. (British.)

"The Territorial Machine Gunner." Printed for private circulation, November 23; 16 pp. A memorandum for the guidance of the section 5th Battalion East Surrey Regiment.

MAIRETET, CAPTAIN. (French.)

"The Infantry Machine Gun: its History and Tactical Employment." Chapelot, Paris; 100 pp., illustrations. Lavau, vol. ii., p. 725.

MEURON, LIEUT.-COLONEL E. DE. (Swiss.)

"The Transport of Machine Guns in Winter." *R.M.S.*, December, p. 1004.

MONDRAGON, GENERAL. (Mexican.)

"The Mexican Automatic Rifle." One diagram. Paris, *R. d'A.*, May.

PAINVIN, MAJOR. (French.)

"Spanish Army Provisional Regulations for the Employment of Machine Guns." Paris, *L.R.I.*, November.

PAINVIN, MAJOR. (French.)

"Prize Shooting between Russian Machine Gun Sections." Translated from the Russian by Major Painvin. Paris, *L.R.I.*, November 15.

PARKER, LIEUT. (U.S.)

"The Technical Side of Machine Gun Organization." *J.M.S.I.*, November, December.

RICHTER, CAPTAIN F. (Austrian.)

"The Tactics and Employment in Action of Infantry Machine Gun Detachments." Seidel, 8vo., 108 pp. Good headings for subjects.

RUSSO-JAPANESE. (Austrian.)

"Observations and Discussion concerning the Employment of Machine Guns in War, based on the Experiences of the Russo-Japanese Campaign." *O.M.Z.*, Vienna.

RUSSO-JAPANESE. (Austrian.)

"Infantry Machine Guns in the Russo-Japanese Campaign." Vienna, *O.M.Z.*, November, p. 1701.

SCHEMPFF, MAJOR-GENERAL VON. (German.)

"The Armoured Machine Gun Motor Car." Berlin, *J.D.A.M.*

SCHLEYER, COLONEL. (Swiss.)

"Schleyer's Automobile Machine Gun." *R.M.S.*, June.

SCHWARZLOSE. (Swiss.)

"Schwarzlose Machine Gun." *R.M.S.*, April, p. 313.

SIGHTS. (British.)

"Sights for Small-Arms and Machine Guns." *Engineering*, December 3.

SNOW TRANSPORT. (Austrian.)

"The Transport of Machine Guns in Heavy Snow." *O.M.Z.*, September.

TACTICS. (Austrian.)

"The Employment of Machine Guns in the Field." *O.M.Z.*, May; also *J.U.S.I.*, February.

TACTICS. (German.)

"The Question of Machine Guns." Berlin, *K.T.Z.*, p. 193.

VASSEUR, LIEUT. L. W. (French.)

"A Scheme for the Organization of Machine Gun Sections on Horseback for the Algerian Cavalry." *R.C.*, May, June.

VELPRY, CAPTAIN. (French.)

"On Machine Guns." Berger-Levrault. A pamphlet.

VIKTORIN, CAPTAIN H. (Austrian.)

"The Training of Cavalry Machine Gun Detachments." *K.M.*, December, p. 1036.

VILLEMONT, CAPTAIN. (French.)

"The Units for Machine Guns in France and in Foreign Countries." Chapelot, 33 pp. See also Lieut. Boucomont. (*Note.*—This is a good work for the study of the Technique and Tactics of Machine Guns.—F. V. L.)

WITZLEBEN, VON. (Swiss.)

"The Machine Gun Detachments in the Japanese Navy." Frauenfeld, Switzerland. *S.Z.A.G.*, May.

1910.

APPLIN, CAPTAIN R. V. K., D.S.O. (British.)

"Machine Gun Tactics in Our Own and Other Armies." *J.U.S.I.*, No. 383, 1910, pp. 34-65. Five tables. Lecture, October 6, 1909. Chairman, Colonel W. N. Congreve, V.C., M.V.O. The following officers took part in the discussion:—Colonel H. de B. de Lisle, C.B., D.S.O., Major E. Shearman, Captain C. A. L. Yate, Captain F. V. Longstaff, T. Miller Maguire, M.A., LL.D., Lieut. H. I. Money, Colonel Applin, and Major Mansell. (*Note.*—This is the best lecture ever given in the R.U.S.I. on Machine Gun Tactics. After stating his case most thoroughly and carefully, the lecturer boldly advocates the training of all the Machine Gunners in a brigade under a selected field

Major R. V. K. Applin, D.S.O.

Author of "Machine-Gun Tactics."

Face page 294.

officer, which officer would command them on service. "We can scarcely expect to obtain a high standard of tactical training or organized bodies capable of manœuvring under fire and combined effort from the regimental subaltern and his two guns left absolutely to his own resources. The best, and nothing but the best, is essential to the successful employment of Machine Guns in war, and the necessity for obtaining the very best officers as section commanders is so great that I am inclined to doubt the utility of having Machine Guns at all if they are not commanded and handled by those who are in every way expert in their use." He lays down the following requirements for a Machine Gun: (1) To fire up to 400 rounds a minute with accuracy, even prolonged, continuous firing; (2) to accompany cavalry and infantry always, and be able to go into action quickly; (3) a firm mounting that can be used kneeling, sitting, or prone; (4) gun and mounting light enough for one man to carry for a considerable distance, and able to be dragged by one man crawling; (5) to be always ready for action, and able to fire when limbered up; (6) to be simple, strong, and durable. In regard to "FIRE POWER, we may say it is equal . . . in fire effect to a half-company (fifty men) of infantry, while it occupies 5 feet of frontage instead of 50 yards. Its vulnerability depends on its mounting, which should be as low as possible. . . . The BEATEN ZONE is perhaps the most important factor in obtaining EFFECTIVE FIRE. MACHINE GUN FIRE is always collective and concentrated, unless deliberately dispersed by the firer; while rifle fire is always individual and dispersed, unless specially controlled by fire discipline under a leader." Before deciding on its (Machine Gun) place in battle, he reviews what has been done in this matter by other armies. He suggests a new method of firing—deliberate rate of 70 to 120 per minute. This imitates rapid rifle fire, and so the presence of a Machine Gun is not disclosed; ". . . in fire action more powerful than infantry. . . ." *Captain Yate :* "If we want to get the maximum efficiency from Machine Guns, I think we must give the Machine Gun officers instructions at the beginning of the day in the same manner as to other officers commanding detached forces and after that leave them a perfectly free hand. This presupposes considerable knowledge and skill on their part, but with PROPER ORGANIZATION and TRAINING we shall be able to insure that they have it." *Major Mansell :* "One instance of well-instructed detachments comes to us from the battlefield of Mukden, where sixteen Machine Guns, used eight at a time, repelled seven fierce Japanese attacks. Each firing lasted only a few minutes, and during that time the eight guns not firing were being overhauled, cleaned, oiled, etc., by the detachments. The sixteen guns fired altogether nearly 200,000 rounds, and at the end of the day every gun was in an excellent condition." In regard to the suitability of pack transport, "In one instance last summer a mule loaded with a Maxim rolled down 250 yards of a very steep side of a hill in the Alps. Only one of the sights was slightly bent, otherwise the gun was quite serviceable. . . ." *Chairman* : " . . . Although we have had the gun longer than any other people in the world—we have had it now for twenty years—we have fallen behind foreign nations in every respect, both in thought, in mechanical devices, and in recognizing the absolute necessity for making the workers SPECIALISTS. The foreigners have given patient thought to the tactics, . . . and have settled a definite policy with regard to it. They (the foreigners) have adopted a very light gun, under 20 pounds, . . . telescopic sights; and the French have got some sort of attachment, which is a very secret thing, by which they can control the rate of fire—a question of very great importance. Finally, they have recognized that it takes three years to make a Machine Gunner. . . . The foreigners, who do not make changes or innovations without very good reason, are generally adopting pack transport for Machine Guns. . . . They think mobility is increased. I think myself, until we do get that brigade gun officer, it will be very difficult to get any improvement. In regard to overhead fire, it is a question of either hearing your own bullets or feeling those of the enemy. . . . The effectual handling of Machine Guns demands hard study and patient experiment."—F. V. L.)

APPLIN, CAPTAIN R. V. K., D.S.O.

" Machine Gun Tactics." London, H. Rees. 265 pp., 8vo. There are 9 plates and diagrams. (*Note*.—This is a very sound study of Tactics, and quite prophetic in many features of the present war. It is most carefully written, and includes many quotations from sound military authorities, so that the reader must study it carefully in order to get all there is to be got out of the book. It contains nine chapters, and covers the whole subject very thoroughly. The chapter on Employment with Infantry, though written six years ago, is still applicable as sound tactics in the great world war. Captain Applin evolved the true Machine Gun tactics in that year, when all armies were engaged in working out the same subject. In some armies, the General Staff were sympathetic with the workers, while in others they were not. I could give many quotations from the above chapter, but a few must suffice: " It has taken nearly forty years to convince the world that tactics are as necessary for the effective employment of Machine Guns as for infantry. If the enemy's artillery is not silenced, or at least dominated, by that of the attackers, it will be necessary to conceal the Machine Guns from view, and provide cover from fire. That Machine Guns, well-concealed and provided with cover (pits), are able to remain in action under artillery fire for considerable periods, is proved by instances in the Russo-Japanese War. Covering fire from the flanks is more likely to be effective than frontal fire, because it will enfilade entrenchments, reach men behind natural cover, and have a more disconcerting moral effect than frontal fire. It will not always be possible to find suitable fire positions on the flanks for Machine Guns at this stage of the attack, and it will then be necessary to place them in rear of the attacking infantry, and fire over their heads. Direct fire should only be used when it is impossible to use indirect fire, on account of the nature of the ground in the vicinity. Indirect fire from the reverse slope of a hill has the advantage of concealing the guns, and rendering them immune from artillery fire. When the covering fire of Machine Guns is no longer considered necessary, they should be withdrawn, and concentrated in batteries in rear of the reserve, or in such other convenient position as the General Officer Commanding may direct." A brigade or divisional Machine Gun officer will find in this book a solution for many of the tactical problems he comes up against. His greatest difficulty in obtaining decisive results against the enemy will be lack of training in his officers, N.C.O.'s, and men. In using artificial cover Captain Applin says: ". . . The necessity for the most thorough peace training, and constant practice in all the details of bringing guns into action, making cover, taking up alternative positions, retiring under mutual support, etc., cannot be too strongly urged on section commanders." The chapter on Employment in Fortress Warfare is very applicable to the trench warfare in France and Flanders. He says: "The Machine Guns of a fortress should be divided into two classes—the stationary and the mobile guns. To the former will be allotted the defence of distinct portions of the permanent works, and they will be provided with cone and parapet mountings, the former being fixed, and the latter capable of being moved within the work to which it is allotted. The mobile guns should be mounted on a light tripod, and carried in a low-wheeled handcart, or they may be mounted on a very light two-wheeled carriage capable of being drawn by one man, and having wheels of small track, which can move over the narrowest roads in the fortress. These mobile guns should be . . . for use in repelling assaults and making counter-attacks. The mobile Machine Guns of the fortress will be used on the advanced line of defence with the mobile troops, and should be divided into two—those allotted to the outposts, and those allotted to the local reserve. The detachments of guns on outpost duty at night should be told off into three watches of two men each, whose duty it will be to remain with the gun in readiness for instant action. The Japanese acknowledge the immense value of Machine Guns to the defence. The searchlights are stationary, they say, and the ground round Port Arthur is broken, so that they can avoid them; but the Machine Guns can be moved about anywhere, and can easily be shifted from place to place by a couple of men." One can only regret that hundreds of officers have not been in training since 1910 on this book of tactics.—F. V. L.)

BUSCHEK, MAJOR-GENERAL W. (Austrian.)

"Musketry Exercises and Field Firing." Seidel, 152 pp.

CAVALRY. (Austrian.)

"How to Engage Hostile and to Protect One's Own Machine Guns in Cavalry Engagements." Vienna, *K.M.*, June.

"Cavalry Machine Gun Detachments." Vienna, *K.M.*, September.

FLECK, CAPTAIN VON A. (German.)

"The Newest Machine Guns." Mittler, Berlin; 140 pp., 24 plates. This book is a continuation of "Machine Guns from a Technical and Tactical Point of View." The illustrations are good.

GREENER, W. W. (British.)

"The Gun and its Development." Ninth edition. London: Cassell and Co. 804 pp. Numerous illustrations and tables. (*Note.*—It is very clearly and simply written. There are chapters on Repeating Rifles, Military Magazine Rifles, Automatic Rifles, and many reports of shooting trials of rifles.—F. V. L.)

HASLINGEN, LIEUT.-GENERAL G. VON. (German.)

"Training for War." This includes chapters on Infantry Machine Guns, Night Operations, and Night Fighting. Berlin, *J.D.A.M.*

JACKSON, LIEUT. V. A. (British.)

"Machine Guns and their Tactical Uses." London, Forster Groom and Co.; 68 pp., 8vo. Major-General Sir H. Rawlinson has written the preface to this little book, in which he discusses briefly the employment of Machine Guns dispersed and in batteries, favouring the former use.

KIESLING, CAPTAIN H. VON. (German.)

"Operation Orders." Translated from the German by the General Staff, British War Office. London; 189 pp., 2 maps. (*Note.*—The orders to a Machine Gun Company are dealt with on pp. 18 to 21. This is a good example of the German idea of combination of Machine Guns with other arms.—F. V. L.) In German the first edition was published at Munich in 1907, and the second edition at Neuburg in 1910.

KRETZSCHMAR, LIEUT. (German.)

"Practice with Machine Guns." Oldenburg, Stalling; 56 pp. (*Note.*—This book contains instructions, drawn from the author's own experience, for the practical application of the teaching of the German Regulations for Machine Gun Companies.—F. V. L.)

MADSEN. (Dutch.)

"Madsen Machine Gun in Foreign Countries." *D.M.S.*, p. 36.

MIORKA, CAPTAIN VON. (Austrian.)
" Machine Guns." *C.J.*, October. This is a translation.

REBOUL, COMMANDANT. (French.)
" The Mitrailleuse in 1870." Paris, Chapelot; 8vo., 163 pp.

SCHARFENORT, CAPTAIN VON. (German.)

" Book of Reference on Military Literature." Mittler, Berlin; 4to., 35s. (*Note.*—This encyclopædia enables the reader to see at a glance what works have been published, and what articles have been written, on any military subject. The subjects are not arranged in alphabetical order, but are grouped under certain general headings, which are again subdivided into minor headings. The chief headings are—

I.—Warfare. The Command of an Army and of Troops.
II.—The History of the Armies of Certain Countries.
III.—Biographies and Memoirs.
IV.—Military History.

SCHIMAK, LIEUT. VON R. (Austrian.)

" Machine Guns in the Infantry Fight: a Few Tactical Schemes worked out with Sketches." *O.M.Z.*, June, p. 885.

SCHULZ, MAJOR. (German.)

" Handbook of Instruction and Duties for Machine Gun Batteries and Machine Gun Companies." Joint author with Captain Roeschke. Berlin, Vossiscke; 212 pp., 8vo. Divided into four parts.

ULLRICH, RICHARD M. (German.)

" The Russian Army under Fire during the War 1904-5." Seidel.

VICHIER-GUERRE, LIEUT. (French.)

" The Tactical Employment of Infantry Machine Guns." Paris *L.R.I.*, March 15.

VIKTORIN, CAPTAIN H. (Austrian.)

" Cavalry Machine Guns on Pack Animals or on Carriages." *K.M.*, November.

WHITE, SERGEANT B. G. (British.)

" Maxim Gun: Instructional Manual for the Mechanism." Forster Groom, 40 pp.

WILSON, MAJOR C. H. (British.)

" Massing of Guns Past and Present." *U.S.M.*, April.
" Machine Guns from an Artillery Point of View." *U.S.M.*, August.

1911.

A. H., Major. (French.)

" Tactical Employment of Machine Guns." *J.S.M.*, February 1.

Balck, Colonel. (German.)

" Tactics." Translated from the German by Lieut. W. Krueger, U.S. Infantry, and published by Rees. First volume. (*Note.*—This contains a good exposition of the current German ideas on Machine Guns.—F. V. L.)

Campbell, Lieut.-Colonel J., D.S.O.

" Fire Action." *A.M.S.* Lecture on March 14, 1911. No. cxii., 15 pp. This was No. 4 of a series of lectures arranged by the General Officer Commanding the Second Division (Major-General H. M. Lawson, C.B.). The following officers took part in the discussion:—Major N. R. McMahon, D.S.O., Captain R. Cockburn, Captain W. S. M. Wetherell, Major C. A. Wilding, Brigadier-General H. K. Jackson, D.S.O., Colonel H. S. Horne, R.A., Brigadier-General F. I. Maxse, C.V.O., C.B., D.S.O., Brigadier-General C. R. Simpson, and Major-General S. H. Lomax. (*Note.*—This lecture is very important as emphasizing the true combination for Machine Gun fire with rapid rifle fire, and that the whole vast subject comes together under fire action; " . . . that the greatest number of troops should advance that the covering fire will allow of. . . . Well-handled Machine Guns should be prepared to assist infantry forward; . . . the advance must be from fire position to fire position. It shows clearly how necessary a high standard and continuous practice in such matters as visual training, quick and consistent aiming, trigger pressing, and rapid loading are, if we wish to obtain a sufficient volume of effective fire on service." *Major McMahon :* " . . . There can really be only one rate of fire—that is, the best rate of every man for combining rapidity with accuracy." *Captain Cockburn :* " . . . But all infantry regiments had a means of using covering fire which was generally neglected, and that was the Machine Gun." *Captain Wetherell :* " One would never, he hoped, try to send troops against fifty Machine Guns massed, and he asked, if one would hesitate to send troops against fifty Machine Guns massed, then at what stage of improvement of the rifle would they modify their tactics, and stop sending troops against it ? Manœuvre, followed by digging in, together with intense rifle and shell fire, would win fights in the future." *Lecturer in reply :* " As regards Captain Cockburn's point about the overhead fire of Machine Guns, that was a question that foreigners were paying great attention to now."—F. V. L.)

Tisch. (Austria.)

"Results of Field Firing Practice of Cavalry and Infantry Machine Gun Detachments." Seidel.

Viktorin, Captain H. (Austria.)

"Machine Guns in the Russo-Japanese War, and Personal Experiences in the Cavalry." Seidel.

1912.

Cordier, Captain. (French.)

"Automatic Weapons." Chapelot, 9 chapters, 135 pp. (*Note.*—This is a most useful book for students. It gives simple diagrams which explain the mechanics of automatic machine guns, rifles, and pistols, —F. V. L.)

Dupeyre, Lieut. (French.)

"Our Machine Guns; what they are and what one may expect of them." Berger-Levrault, 99 pp. (*Note.*—This is an appreciation up to date of the material, training, and tactics, in the most important armies of the world. The author makes frequent use of quotations from both official and unofficial books.—F. V. L.)

Melczer. (Austria.)

"Main Points for Training in the Service of Machine Guns." Seidel.

Merkatz. (German.)

"The New Machine Gun and its Drill." Seidel.

Schoenlaub. (French.)

"Regulations for the German Machine Gun Service." (Groups and companies.) Translated from the German by Captain Schoenlaub. Lavauzelle, 128 pp., 13 engravings in text. This includes the amendments up to 1909.

Seidel. (Austrian.)

"Target Practice for Infantry and for Infantry Machine Gun Detachments." Seidel.

1913.

Ebner, Lieut. (French.)

"Machine Gun Sections" (concluded). *L.R.I.*, June 15, p. 419, per *A.R.*, October.

Fleck, Major. (German.)

"Machine Guns, the Newest Features." Mittler. (*Note.*—This is a supplement to Fleck's two previous books on this subject.—F. V. L.)

KRIEGER, A. (German.)

"Machine Gun Companies in Action." Oldenburg.

LEWIS. (French.)

"The Lewis Machine Gun." Paris, *L.R.I.*, July 15, p. 85, per *A.R.*, October.

MERKATZ, CAPTAIN F. VON. (German.)

"Machine Guns: Their Technique, Ballistics, and Application." Mittler, Berlin. A handbook for instruction; 3 drawings.

OSTERMEYER. (German.)

"The Fight of the Machine Gun Company." Berlin.

RUSSIA. (Russian.)

"A Review of Four Official Manuals for Machine Guns in the Russian Army." *A.R.*, October, pp. 635-638. The books are as follows:

Infantry Machine Gun Detachment Training, 162 pp.
Cavalry ,, ,, ,, ,, 124 pp.
Instructions for Musketry Training with Machine Guns, 199 pp.
Description of the Machine Gun Equipment. Part I., 108 pp.; Part II., 172 pp.

SKODA. (Austrian.)

"The Skoda Steel Works, Pilsen, Bohemia." *Engineer*, January 17, pp. 60-64. (*Note.*—The Skoda Works started in 1859. They are now by far the largest steel works in Central Europe. The head offices are in Vienna.—F. V. L.)

STOGER-STEINER. (Austrian.)

"On the Tactical Use and Leading of Infantry Machine Gun Detachments in Action." Seidel.

1914.

BALCK, COLONEL. (German.)

"Tactics." Vol. II.: Cavalry and Artillery. Translated from the German by Lieut. W. Krueger, 23rd U.S. Infantry. London, H. Rees.

BOSTOCK, SERGT.-MAJOR. (British.)

"The Machine Gunner's Handbook, including the Vickers Light Gun." London, W. H. Smith and Son. October, 270 pp., 20 plates. (*Note.*—The preface states: "These notes show, in some measure, the system followed at the School of Musketry at Hythe, in giving effect to the instructions laid down in the official textbooks on Machine Guns, and are compiled with a view to assisting instructors in framing a correct sequence of instruction, and with the desire to increase interest in what is undoubtedly the weapon of the future." This is a most valuable book of reference.—F.V.L.)

COLT. (U.S.)

"The Colt Automatic Gun." Colt's Patent Firearms Manufacturing Company, Hartford, Connecticut, U.S.A. A trade handbook published by the Company. Model 1914, 28 pp., 16 illustrations.

IRONSIDE, LIEUT. H. A. (British.)

"The Machine Gun: Its Drill, Signals, and Control.' London, H. Rees; 30 pp., 7 diagrams.

LEWIS. (British.)

"Handbook of the Lewis Automatic Machine Gun, Model 1914." Published by Armes Automatiques Lewis, Anvers, Belgique. London office, 27 Pall Mall, S.W. Manufactured by the Birmingham Small-Arms Company, Ltd. 24 pp., 5 plates and 5 figures.

VANDERPUTTE, LIEUT. (Belgian.)

"The Belgian Military Dog employed to draw Maxim Machine Guns." Brussels, 79 pp., 9 chapters. Guyot Frères. (*Note.*—This book deals with the selection, care, and training, of dogs for drawing Machine Guns. To use a new word, it is on "Dogmastership."—F. V. L.)

1915.

APPLIN, MAJOR R. V. K., D.S.O. (British.)

"Machine Gun Tactics." Third edition, revised. London, H. Rees. 194 pp., 1 map, several diagrams, and one sketch. Extract from Preface: "The great demand for this book, due to the war, compelled the publisher to issue the second edition without waiting for the author's corrections. A third edition is now called for, and the author has been able to bring the book up to date in all essentials to conform to the latest teachings of the official textbooks. Many changes have taken place in the organization and training of Machine Gun detachments since the author wrote the first edition, notably the brigading of guns, as therein suggested, so that it is with real satisfaction that he finds the tactical principles advocated have been so fully confirmed by the present war that it has not been necessary to alter this part of the book in any essential detail." Dated at School of Musketry, Satara, India, July 21, 1915. (*Note.*—There are many minor alterations in accordance with Field Service Regulations, several footnotes from experience of the present war. The chief additions are: (1) Indirect fire for Machine Guns by graticule to determine the position of the guns with reference to target, so that the trajectory will clear the rising ground between, which is simplified by means of the SATARA DIRECTOR; (2) practical notes on the care and adjustment of gun; (3) drawing of and method of using the SPIKE mounting. This is a most valuable book for constant study.—F. V. L.)

ALCANTARA, CAPTAIN M. (Brazilian.)

"Machine Guns." Rio Janeiro; two volumes. Vol. I.: Theory of Machine Guns. Comprises—Origin and Evolution of Machine Guns; General Problem of the Machine Gun; Modern Conception of the Machine Gun; Practical Value of Systems and Types; Fire of Past, Present, and Future; Technical Considerations; The Use of Machine Guns in Connection with Other Arms. Vol. II.: Description of Types. Per the *International Military Digest*, U.S.A.

BÉNET-MERCIER. (U.S.)

"Machine Gun Experimental Firing." *Army and Navy Journal*, May 29, 1915, New York.

BRADBROOKE, LIEUT. C. A. (Canadian.)

"Some Hints to Brigade and Battalion Machine Gun Officers." Hythe, W. S. Paine and Co. 16 pp. (*Note.*—These notes are specially for the present trench warfare in France, and all the information has been gleaned from notes received from the front.—F. V. L.)

CHARTERIS, CAPTAIN N. K. (British.)

"Some Lectures and Notes on Machine Guns." London, W. H. Smith and Son. 106 pp., illustrations and diagrams. *Contents:* I. History, Organization, and Training.—II. Characteristics.—III. Amplified Notes on Parts I. and II.; Table "C" Machine Gun Course.—IV. Fire Direction and Fire Orders.—V. Elementary Training in Tactical Handling.—VI. Useful Hints for Machine Gunners on Active Service.—VII. Examples of the Successful Use of Machine Guns in War. (*Note.*—This is one of the best books for the advanced student. —F. V. L.)

COLT. (French.)

"The American Machine Gun Colt." Paris, Lavauzelle; 37 pp.

CROSSMAN, E. C. (U.S.)

"Machine Guns and their Fire." *U.S.M.*, April, pp. 68-78. (*Note.*—This is a most important article on the higher training and great skill required in the work.—F. V. L.)

HOTCHKISS. (French.)

"Instruction on the Hotchkiss Machine Gun." Lavauzelle, 42 pp. This includes nomenclature, mechanism, stripping, and drill.

JAMES, LIEUT. D. McG.

"Instruction in the Machine Gun." 177 pp. 13 plates, 22 chapters. Forster, Groom and Co. (*Note.*—The matter in this book is sound and well arranged. It is concise, and covers a lot of study, but it does not give much about GROUND OR TACTICS in the BATTLE.—F. V. L.)

JOHNSTONE, CAPTAIN D. J. (British.)

" Handbook of the Colt Gun." Hythe, W. S. Paine. 1s. 6d. 28 pp., diagrams. This book is written for the officers of the Canadian Training Division at Shorncliffe.

MAXIM. (German.)

" The Maxim Machine Gun." French Translation of the German Regulations. Lavauzelle. Second edition, 59 pp., illustrations.

VICKERS. (British.)

" Guide for the ·303-Inch Vickers Machine Gun: its Mechanism and Drill, with Questions and Answers." Aldershot, Gale and Polden, 99 pp., 6d. Plates and tables.

VICKERS. (French.)

" The English Vickers Machine Gun." Paris, Lavauzelle. 38 pp., illustrations. In French.

1916.

ROUSE, CAPTAIN S. (British.)

" Practical Notes for Machine Gun Drill and Training." London: Forster Groom. 82 pp. To be read in conjunction with official handbooks. (*Note.*—The whole scheme of this little book appears to have been clearly worked out. By far the best part of the book is the " Elementary Drill without Transport."—F. V. L.)

" SIMPLEX."

" Instruction on the Lewis Automatic Machine Gun." London: Forster Groom. 143 pp. Several plates and diagrams. Part I., Mechanism and Drill; Part II., The Handling of the Gun. (*Note.*—This is an excellent little work well worthy of the closest study. The title of Part II. should be " Minor Tactics of Machine Guns." Its weakness is the omission of a chapter on the " Elementary Study of Ground," but this is counterbalanced a little by the good chapter on " Range-Taking."—F. V .L.)

APPENDIX II

BRITISH PATENT OFFICE RECORDS

ABRIDGMENT OF SPECIFICATIONS

CLASS 92, ORDNANCE AND MACHINE GUNS

THE following is a list of most of the patents concerning machine guns, and is intended as a guide to the student. The Patent Office is between Chancery Lane and Staples Inn. Each patent is designated by its year, and has its serial number for that year.

The list given is in two parts—

(i.) Chronological list by serial numbers. (Those numbers starred are in the alphabetical list.)
(ii.) Alphabetical list under name of patentee.

NOTE.—The total entries are, List (i.), 79; List (ii.), 68 For further information on this subject look up "Lake" in Bibliography under year 1895.

LIST (i.)

1865	790.
1873	1,739.*
1878	2,735, 3,017,* 3,678.*
1879	1,317, 4,454.*
1881	5,436.
1883	3,493.
1884	606,* 3,844,* 9,407,* 13,113.*
1885	1,307,* 8,281,* 14,049.*
1890	483,* 6,591,* 17,857,* 16,939.*

1891	... ...	16,081,* 7,137.*
1892	... ...	7,156.*
1894	... ...	16,260,* 20,627.*
1895	... ...	5,864.*
1896	... ...	5,426.*
1897	... ...	11,713.*
1898	... ...	**88**, 8,027,* 19,714.*
1899	... ...	23,271.*
1900	... ...	14,921,* 17,639,* 20,865,* 21,630,* 21,743.*
1903	... ...	14,310.*
1906	... ...	4,389,* 7,161.*
1907	... ...	6,949.*
1908	... ...	3,092,* 5,141,* 6,680,* 6,845,* 10,312,* **10,418**, 13,538,* **14,310**, 14,966,* 16,370,* 18,520,* 18,843,* 18,849,* 25,269.*
1910	... ...	865,* 24,232,* **24,255**, 24,258.*
1911	... ...	11,233,* 15,660,* 15,661,* 15,663.*
1912	... ...	**1,138**, 1,675,* 3,050,* 5,013,* 9,914,* 11,475.*
1913	... ...	**1,032**, 3,412,* 2,464,* 3,559,* 6,723,* 6,724,* 18,276,* 20,009.*
1914	... ...	1,688,* 3,571.*

LIST (ii.)

ALPHABETICAL LIST UNDER NAME OF PATENTEE

ACLAND. 1891. No. 16,081.
"Improvements in Carriages or Mountings for Light Guns."
F. E. D. Acland, late Captain R.A. 32, Victoria Street, Westminster.

ARMSTRONG. 1913. No. 20,009.
"Improvements in Automatic Guns."
Sir W. G. Armstrong; Whitworth and Co., manufacturing engineers; A. G. Hadcock, late R.A.; and G. Forster, draughtsman—all of Elswick Works, Newcastle-on-Tyne.

AUGEZD. 1890. No. 16,939.
"Improvements in Machine Guns."
A. O. F. von Augezd, an officer in Austro-Hungarian Army, of 111 Strohgasse, Vienna, Austria.

AUGEZD. 1892. No. 7137.
"Improvements in Machine Guns."
A. O. F. von Augezd, an officer in Imperial Army, of 111 Strohgasse, Vienna, Austria.

AUGEZD. 1899. No. 23,271.
"Fire-Arm for Firing Single Shots and Automatic Continuous Fire."
A. Baron Odkolek von Augezd, Imperial and Royal Cavalry Instructor, of 1 Opernring, Vienna.

AUGEZD. 1906. No. 4389.
"Improvements in Automatic Machine Guns."
Baron A. Okolek von Augezd, gentleman, of 5 Maximillanstrasse, Vienna, Austria.

BERGMAN. 1890. No. 483.
"Improvements in and relating to Breech-Loading Small-Arms and Machine Guns."
O. W. Bergman, Lieutenant Royal Swedish Artillery, of Gothenburg, Sweden.

BERGMAN. 1890. No. 17,857.
"Improvements in Breech-Loading Fire-Arms."
O. W. Bergman, Lieutenant Royal Swedish Artillery, of Gothenburg, Sweden.

BJORGUM. 1911. No. 15,660.
"Improvements in Automatic Fire-Arms."
N. Björgum, inventor, of Asker, near Christiana, Norway.

BJÖRGUM. 1911. No. 15,661.
"Cartridge Feeding Device for Automatic Fire-Arms."
N. Björgum, inventor, of Asker, near Christiana, Norway.

BJÖRGUM. 1911. No. 15,663.
"Cartridge Belt for Automatic Fire-Arms."
N. Björgum, inventor, of Asker, near Christiana, Norway.

BOULT. 1900. No. 21,743.
"Improvements in or relating to Machine Guns."
A. J. Boult, chartered patent agent, of 111, Hatton Garden, London. A communication from abroad by G. Perino, of Rome.

BOULT. 1908. No. 18,849.
"Improvements in or relating to Machine Guns."
A. J. Boult, chartered patent agent, of 111, Hatton Garden, London. A communication from abroad by (F.I.A.T.) Fabrica Italiana Automobili-Torino, Societa Anonima, Turin, Italy.

BROWNING. 1900. No. 14,921.
"Automatic Gun."
J. M. Browning, manufacturer, of 505, 27th Street, Ogden, Utah, U.S.A.

CALDWELL. 1913. No. 18,276.
"An Improved Manually Operated Machine Gun."
T. F. Caldwell, mechanical engineer, of 6, Church Street, Richmond, Victoria, Australia.

DAWSON. 1908. No. 13,538.
"Improvements relating to Automatic Guns."
A. T. Dawson, Lieutenant R.N., director and superintendent of ordnance works of Vickers, Sons and Maxim, 32, Victoria Street, Westminster. A communication from abroad by the Deutsche Waffen und Munitions Fabriken of Berlin.

DAWSON. 1908. No. 16,370.

"Improvements relating to Automatic and Similar Machine Guns."

A. T. Dawson, Lieutenant R.N., director and superintendent of ordnance works of Vickers, Sons and Maxim, 32, Victoria Street, Westminster. A communication from abroad by the Deutsche Waffen und Munitions Fabriken of Berlin.

DAWSON. 1898. No. 19,714.

"An Improved Automatic Machine Gun."

A. T. Dawson, late Lieutenant R.N., and L. Silverman, engineer, both of 32, Victoria Street, London.

GOERZ AKTIENGESELLSCHAFT. 1914. No. 3571.

"Improvements in Telescopic Sights for Machine Guns with Protecting Shields."

Optische Anstalt C. P. Goerz Aktiengesellschaft, of Berlin-Friednau, Germany.

HOTCHKISS. 1879. No. 4454.

"Revolving Cannon, etc."

B. B. Hotchkiss, engineer, of 38, Southampton Buildings, Chancery Lane, London.

HOTCHKISS. 1896. No. 5426.

"Improvements in Machine Guns."

The Hotchkiss Ordnance Company, Ltd., of 49, Parliament Street, London, England. A communication from abroad by L. V. Benét, a citizen of the U.S., and H. A. Mercié, a citizen of the Republic of France, inventors, both residing in Paris, France.

HOTCHKISS. 1908. No. 5141.

"Improvements in Cartridge Feed Mechanism for Automatic Guns."

The Hotchkiss Ordnance Company, Ltd., of 25, Victoria Street, London, England. A communication from abroad by L. V. Benét and H. A. Mercié, artillery engineers, both of Paris, France.

IMRAY. 1900. No. 20,865.

"An Improvement in the Cover Tubes and Training Pivots of Machine Guns."

O. Imray, chartered patent agent, Birkbeck Bank Chambers, Southampton Buildings, London. A communication from abroad by the Deutsche Waffen und Munitions Fabriken, of Dorotheenstrasse 43-44, Berlin.

IMRAY. 1900. No. 21,630.

"An Improved Gun Carriage."

O. Imray, chartered patent agent, Birkbeck Bank Chambers, Southampton Buildings, London. A communication from abroad by the Deutsche Waffen und Munitions Fabriken, of Dorotheenstrasse 43-44, Berlin.

KRINNINGER. 1913. No. 3412.

"Improvements in Gun Mountings."

G. Krinninger, First-Lieutenant, of Niederdorf, Tyrol, Austria.

LAKE. 1890. No. 6591.
"Improvements relating to Automatic Guns."
H. H. Lake, of Haseltine Lake and Co., patent agents, Southampton Buildings, Middlesex. A communication from abroad by H. S. Maxim, engineer, of Crayford Works, Kent, residing in Madrid, Spain.

MARKS. 1912. No. 9914.
"Improvements in Gas Operated Guns."
E. C. R. Marks, consulting engineer, of 57, Lincoln's Inn Fields, London. A communication from abroad by the Automatic Arms Company, of 504, D. S. Morgan Buildings, Buffalo, New York, U.S.A.

MAXIM. 1884. No. 13,113.
"Improvements in and relating to Machine and other Guns."
H. S. Maxim, engineer, of 57D, Hatton Garden, Middlesex.

MAXIM. 1884. No. 606.
"Improvements in and relating to Machine Guns and other Fire-Arms."
H. S. Maxim, mechanical engineer, of Cannon Street, London.

MAXIM. 1884. No. 3844.
"Improvements in and relating to Machine or Battery Guns."
H. S. Maxim, engineer, of 59D, Hatton Garden, Middlesex.

MAXIM. 1884. No. 9407.
"Improved Feed-Apparatus for Machine or Battery Guns or other Fire-Arms."
H. S. Maxim, engineer, of 59D, Hatton Garden, Middlesex.

MAXIM. 1885. No. 1307.
"Improvements in and relating to Machine or Battery Guns and other Fire-Arms."
H. S. Maxim, engineer, of 57D, Hatton Garden, Middlesex.

MAXIM. 1885. No. 8281.
"Improvements in Machine and other Guns."
H. S. Maxim, engineer, of 57D, Hatton Garden, Middlesex.

MAXIM. 1885. No. 14,047.
"Improvements in Machine and other Guns and Pistols, and in Projectiles Therefor."
H. S. Maxim, engineer, of 57D, Hatton Garden, Middlesex.

MAXIM. 1892. No. 7156.
"Improvements in Automatic Guns."
H. S. Maxim and J. Silverman, engineers, both of Crayford Works, Kent.

MAXIM. 1894. No. 16,260.
"Improvements in and relating to Automatic and Machine Guns, and their Stands and Supports."
H. S. Maxim and J. Silverman, engineers, both of Crayford Works, Kent.

MAXIM. 1894. No. 20,627.
"Improvements in and relating to Automatic Guns."
H. S. Maxim, engineer, of Baldwyn's Park, Bexley, Kent.

MAXIM. 1895. No. 5864.
"Improvements in and relating to Automatic Guns."
H. S. Maxim, engineer, of Baldwyn's Park, Bexley, Kent.

MAXIM. 1903. No. 14,310.
"Improvements in Devices for Lessening the Sound of Discharge of Guns."
H. P. Maxim, engineer, of 550, Prospect Avenue, Hartford, Connecticut, U.S.A.

MAXIM. 1908. No. 6680.
"An Improved Device for Lessening the Sound of Discharge of Guns."
Sir H. S. Maxim, Chevalier of the Legion of Honour, civil, mechanical, and electrical engineer, of Thurlow Lodge, Norwood West, Surrey.

MAXIM. 1908. No. 6845.
"Improved Device for Lessening the Sound of Discharge of Fire-arms."
H. P. Maxim, engineer, of 550, Prospect Avenue, Hartford, Connecticut, U.S.A.

MAXIM. 1908. No. 25,269.
"Improvements in Devices for Lessening the Sound of Discharge of Guns."
H. P. Maxim, engineer, of 550, Prospect Avenue, Hartford, Connecticut, U.S.A.

MORGAN-BROWN. 1878. No. 3017.
"Revolving-Cannons."
W. Morgan-Brown, of Brandon, Morgan-Brown, engineers and patent agents, 38, Southampton Buildings, London, and 1 Rue Laffitte, Paris. A communication from abroad by B. B. Hotchkiss, of 1 Rue Laffitte, Paris.

NORDENFELT. 1873. No. 1739.
"Battery Gun."
T. Nordenfelt, of St. Swithin's Lane, London.
Nordenfelt's Disclaimer and Memorandum of Alteration. A communication from abroad by H. Palmerantz, civil engineer, and J. T Winborg, manufacturer, both of Stockholm; and E. Unge, of Motala, Sweden.

NORDENFELT. 1878. No. 3678.
"Battery Guns."
T. Nordenfelt, civil engineer, of 1, St. Swithin's Lane, London.
A communication from abroad by H. Palmerantz, civil engineer, of Stockholm, Sweden.

ODKOLEK. 1897. No. 11,713.
"Improved Breech-Closing Mechanism for Machine Guns."
A. Baron d'Odkolek, Captain Imperial and Royal Army (retired) of Fichtenhof, near Klagerfurt, Carinthia, Austria.

PERINO. 1907. No. 6949.
"Improvements in Feeding Automatic Guns."
G. Perino, chief technician of the artillery, Rome, Italy.

PERINO. 1912. No. 11,475.
"Improvements in Machine Guns."
G. Perino, mechanist, of Turin, Italy.

SCHWARZLOSE. 1910. No. 865.
"Improvements in Automatic Fire-Arms."
A. W. Schwarzlose, manufacturer, of Charlottenburg, Germany.

SCHWARZLOSE. 1912. No. 3050.
"Breech Mechanism for Machine Guns."
A. W. Schwarzlose, manufacturer, of Charlottenburg, Germany.

SCHWARZLOSE. 1912. No. 5013.
"Cartridge Frame for Machine Guns."
A. W. Schwarzlose, manufacturer, of 3 Linsenplatz, Charlottenburg, Germany.

SCHWARZLOSE. 1913. No. 2464.
"Breech Mechanism with Knuckle-Joint for Automatic Fire-Arms having Stationary Barrels."
Andreas W. Schwarzlose, engineer, of Charlottenburg, Germany.

TROCHAIN. 1898. No. 8027.
"Improvements in Cartridge Belts for Use in Automatic or Machine Guns."
E. Trochain, of 10 Rue du Château d'Eau, Paris.

VICKERS. 1900. No. 17,639.
"Improvements in Automatic Guns."
A. Vickers, steel manufacturer, and Vickers, Sons and Maxim, Ltd., 32, Victoria Street, Westminster.

VICKERS. 1906. No. 7161.
"Improvements in Automatic Guns."
A. T. Dawson, Lieutenant R.N., director and superintendent of ordnance works, and G. T. Buckham, engineer, both of Vickers Sons and Maxim, Ltd., 32, Victoria Street, Westminster Haseltine Lake and Co., 8, Southampton Buildings, London agents.

VICKERS. 1908. No. 3092.
"Improvements relating to Automatic and Similar Guns."
A. T. Dawson, Lieutenant R.N., director and superintendent of ordnance works, and G. T. Buckham, engineer, inventors, both of Vickers, Sons and Maxim, 32, Victoria Street, Westminster.

VICKERS. 1908. No. 10,312.

"Improvements in and relating to the Elevating Apparatus of Light Guns."

A. Vickers, managing director of Vickers, Sons and Maxim, 32, Victoria Street, Westminster. A communication from abroad by the Deutsche Waffen und Munitions Fabriken, of Berlin.

VICKERS. 1908. No. 14,966.

"Improvements in Tripod and Similar Mountings for Automatic Guns."

A. T. Dawson, Lieutenant R.N., director and superintendent ordnance works, and Carl A. Larsson, engineer, both of Vickers, 32, Victoria Street, Westminster.

VICKERS. 1908. No. 18,520.

"Improvements in the Cartridge Feed Mechanism of Maxim Guns."

Vickers, Sons and Maxim, Ltd., and G. T. Buckham, engineer, both of 32, Victoria Street, Westminster, London.

VICKERS. 1908. No. 18,843.

"Improvements in the Breech Mechanism of Automatic and Similar Guns of Small Calibre."

A. Vickers, managing director of Vickers, Sons and Maxim, 32, Victoria Street, Westminster. A communication from abroad by the Deutsche Waffen und Munitions Fabriken of Berlin.

VICKERS. 1910. No. 24,232.

"Improvements in or relating to Machine Guns."

Sir A. T. Dawson, Knight, Lieutenant (retired) R.N., superintendent of ordnance works, and G. T. Buckham, both of Vickers, Sons and Maxim, Victoria Street, Westminster.

VICKERS. 1910. No. 24,258.

"Improvements in or relating to Automatic Guns."

Sir A. T. Dawson, Knight, Lieutenant (retired) R.N., superintendent of ordnance works, and G. T. Buckham, engineer, both of Vickers, Sons and Maxim, 32, Victoria Street, Westminster.

VICKERS. 1911. No. 11,233.

"Improvements in or relating to Gun Limbers."

Sir A. T. Dawson, Knight, Lieutenant (retired) R.N., superintendent of ordnance works, and G. T. Buckham, engineer, both of Vickers, Sons and Maxim, 32, Victoria Street, Westminster.

VICKERS. 1912. No. 1675.

"Improvements in or relating to Automatic Guns."

By Vickers, Ltd., Vickers House, Broadway, Westminster. A communication from abroad by the Deutsche Waffen und Munitions Fabriken, of Berlin.

VICKERS. 1913. No. 3559.

"Improvements in or relating to Automatic Gun Mountings."

Sir A. T. Dawson, Knight, Lieutenant (retired) R.N., superintendent ordnance works, and G. T. Buckham, engineer, both of Vickers, Ltd., Vickers House, Broadway, Westminster.

VICKERS. 1914. No. 1688.

"Improvements in or relating to Automatic Gun Mountings."

Sir A. T. Dawson, Knight, Lieutenant (retired) R.N., superintendent ordnance works, and G. T. Buckham, engineer, both of Vickers, Ltd., Vickers House, Broadway, Westminster.

VERDUD. 1913. No. 6723.

"Improvements relating to Automatic Rifles."

J. A. y. Verdud, engineer, 31, West View Road, Barrow-in-Furness.

VERDUD. 1913. No. 6724.

"Improvements in Automatic Sporting or the like Guns."

J. A. y. Verdud, engineer, of 31, West View Road, Barrow-in-Furness.

Lieut.-Colonel C. B. Mayne.

Author of "Infantry Fire Tactics" and "The Infantry Weapon and its Use in War."

Face page 315.

APPENDIX III

SELECTED extracts from works on "Infantry Fire Tactics" by the late Lieutenant-Colonel C. B. Mayne, R.E. These are printed here, as the works referred to are now absolutely out of print.—F. V. L.

NOTE.—The italics marked "*" are mine.—F. V. L.

In 1884 Captain C. B. Mayne, R.E., published "Infantry Fire Tactics," about 500 pages, with many tables and diagrams. It was the first standard work on the subject by a British officer. The subject-matter was divided into three main parts: (1) Individual fire and its applicability to short ranges only; appreciation of distances. (2) Collective fire for long ranges; influence of ground on fire; long and short range fire; supply of ammunition; controlled and uncontrolled fire; fire discipline, control and direction. (3) Tactical deductions; musketry instruction; the spade in warfare; repeating rifles.

In 1888 he published a second edition while he was instructor at the Royal Military College, Kingston, Ontario, Canada. It contained a considerable revision and re-grouping of matter, with much fresh information. He says:

"The few pages on Musketry Fire Tactics, to be found in our Infantry Drill Book of 1884, are a deliberate and unacknowledged translation from the French Musketry Regulations, so much so that, where the French have laid down the outer limit of long-range fire as the range for the extreme graduation of their backsight, 1,600 metres, or 1,700 yards nearly, the same limit has been inserted in our Drill Book, although our backsight is only graduated up to 1,400 yards! And all these French statistics for the Gras rifle, after having been dressed up in English units of measure, are headed 'Limits for the Employment of Fire with the Martini-Henry Rifle!'"

He says, just as this book was being printed, the author received a copy of the new official pamphlet, "Infantry Tactics as Influenced by Fire," which is evidently intended to replace the few pages on Musketry Fire Tactics in the Field Exercises of 1884. Much of this pamphlet has been bodily taken from this book, and in it the subdivision of ranges, recommended on p. 266, has been practically adopted. Captain Mayne states on p. 185 that a great portion of chapter xi. (Influence of Ground and Obstacles on the Effects of Infantry Fire. Inclined Fire. Indirect Fire. Night Firing) is taken from the French "Règlement sur l'Instruction du Tir."

On p. 226 he says: "One thing that must strike everyone after a study of the foregoing pages (chapter xi.) is the great importance which must be nowadays attached to *the knowledge of ground and its employment,* in knowing its use offensively and defensively, and its effect on the fire, both in the attack and defence. The value of ground is not absolute; it not only varies with the nature of the arms, with their range and combinations, but it depends also on the actual positions that the troops occupy on it. A knowledge of the ground is no less indispensable for the attack than the defence—here to profit by some strong points, there to avoid them. The ground dictates to the defence the points of resistance and the tactical dispositions; it indicates to the attack the directions in which a bayonet attack has no chance of success, and those where it can succeed——. Tactical dispositions ought to be based on the properties of the ground; an ideal formation on a horizontal ground would be annihilated if it were blindly placed on intersected and varied ground: there does not exist any panacea applicable to all cases."

On p. 231 under Indirect Fire, he says: "By indirect fire is meant any fire directed on objects that are masked from the view of the firers, and which are at some distance in rear of the covering obstacle. Thus the problem of indirect fire reduces itself to this case—namely, to determine the elevation with which one should aim at a visible point, chosen as an auxiliary object, in order that the mean trajectory may pass over the intervening obstacle, and through the centre of the real object."

The late Lord Wolseley thought so highly of this second edition, that in 1905 he presented a copy to the Royal United Service Institution.

In 1903 Lieutenant-Colonel C. B. Mayne, R.E., brought out a third edition of his work, but gave it a different title, "*The Infantry Weapon and its Use in War.*" The book contains about 350 pages, an Index, and such an instructive *Preface* that an extract is given.

"The manuscript of the following pages was sent home in December, 1901, for publication, but delays arising from my being in India have prevented its earlier issue. Since the book was written, 'Infantry Training (Provisional), 1902,' 'Combined Training (Provisional), 1902,' and the 'Musketry Regulations (Provisional), 1903,' have been issued. However, no changes have been made in the original text of the book, but some additions have been made in the form of footnotes. This book has been written in response to the numerous requests that have been received at various times and from various quarters for a revision of my book on 'Infantry Fire Tactics.' But, while correcting the proof sheets, it has occurred to me that I have not laid sufficient stress on the necessity for a close and continuous co-operation of the artillery and infantry arms in battle, since, especially nowadays, each arm requires the assistance of the other arm to enable it to do its work of giving assistance to this other arm. The reader should bear this in mind while perusing these pages. My work on 'Infantry Fire Tactics' was first published in 1884, and an enlarged and revised edition was issued in 1888, nearly fifteen years ago. The book dealt with the use of the single-loading Martini-Henry rifle, with its large-bore bullet, relatively high trajectory, and smoky powder ammunition; and its primary purpose was to plead for the introduction into our Regulations of many important usages and methods of firing that were necessary for ensuring the proper use of that particular rifle in war. The whole of these usages and methods have since then been adopted for some time past, but the large-bore, single-loading, high-trajectory Martini-Henry rifle, with its black-powder ammunition, has now been superseded by a small-bore, flat-trajectory rifle, with magazine attachment and smokeless-powder ammunition. Consequently 'Infantry Fire Tactics' is now practically out of date in many respects, though it has done its work."

"In publishing the present work I have taken up the subject of infantry fire on more general lines than I did in my former book, and I am hopeful that it may prove equally useful to our Army. * *Our campaigns in the Tirah and South Africa have brought many special features to the forefront in connection with the use of the modern rifle in the field. But it seems to me that the exceptional character of these wars has not been fully separated by many writers from the somewhat accidental experiences that our troops gained in them, and consequently that there is, with such writers, a strong tendency to form imperfect and one-sided views on the future use of the rifle from the want of a proper analysis of our experiences, separating their accidents from their substance.* It will be seen I have endeavoured to combat this tendency by making the necessary analyses where needed."

"In such very important matters as the employment of the various aims of the service in the field, I would plead for the official encouragement of the freest and fullest expression of individual professional opinion on all technical matters relating to the Army, *provided that the existing Regulations are, in the meantime, complied with so long as they are in force.* This is a most important and vital matter for the future of our Army, and all the more so with us on account of the undoubted want of touch that exists between our staff and regimental officers. Field-Marshal Earl Roberts, General Sir Redvers Buller, and others have again and again spoken out clearly and strongly about the need that exists for all ranks to think and to be ready to act on their own initiative. But these words of theirs need not ever have been spoken if, in the past, all ranks had been allowed to think and make use of an intelligent initiative. But they were not allowed to do this, and in these respects the Boer War has proved to be a veritable nemesis for the errors that have been forced on regimental officers in the past, and about which they were not permitted to express their views. * *However, personal thinking and the individual initiation of action can only be secured by encouraging men to think and speak and act, under due and liberal control.* But besides *knowledge* there is needed *will-power* to put it into *practice*. Knowledge can be acquired by reading, study, and discussion. And no doubt there are many officers who know what to do under certain circumstances, but who fail in will-power to carry it out. This will-power can be cultivated, or entirely obliterated, by the way in which officers and men are trained in peace time, and it is far better for them to make mistakes—for all learners *must* make mistakes—provided they had some reason for what they did, than for no action to be taken at all for fear of making a mistake. The fear of blame, and the consequent unwillingness to accept responsibility, has had largely to do with the want of initiative in our Army. * *We want in the future both a frank expression of military thought and opinion, and also a generous recognition of the educational value of mistakes made during training.*"

The work contains chapters on: The Initiative; The Infantry Weapon; Long-Range Firing and Range Finding; Invisibility; Cover and Obstacles; Concentrated Collective Firing; Indirect Firing, Rests, Telescopic Sights, Night Firing, and Shields; Methods for ensuring the Rapid Loading of Rifles; Ammunition Supply; Fire Discipline, Control, and Direction; Notes on Musketry Training; Notes on German Musketry Training.

In the chapter on Fire Control, p. 165, there is a most instructive paragraph which is given here: "We can now proceed to consider *the use of the rifle in battle*—namely, the 'direction' of, the 'control' of, and 'the discipline' for, infantry fire. The duty of *directing* the fire falls upon the company leaders and officers senior to them; the duty of controlling the fire falls on the junior officers and senior non-commissioned officers; while the *discipline* required, in order to enable this direction and control to be carried out, falls on the rank and file."

The further remarks on Fire Direction are so much to the point that they are given in full:

"'Fire Direction' consists in determining, in accordance with the governing tactical idea of the battle, (1) the moment for opening the fire; (2) the proper tactical extension of the men or density of firing-line; (3) the advance to shorter ranges; (4) the selection of the objects to be fired on and their allotment to different portions of the firing-line; (5) the back-sight elevations to be used; (6) the observation of the results of the fire; (7) the kind of fire to be used; (8) the amount of ammunition to be expended at each halt; (9) the rapidity of the fire; (10) the strengthening of the firing-line, when and where necessary; (11) the replenishing of the expended ammunition, when and where possible; and (12) the psychological moment for advancing to the bayonet assault."

"These several factors of the art of fire direction are closely inter-related, and cannot be fully separated from one another. However, we have to separate them mentally by analysis, but in practice they require to be combined synthetically before orders are issued. In fact, every military order is the outcome of a synthesis of many and varied considerations."

Still in the same chapter, but p. 193: "The requirements of the modern battlefield demand that every officer and non-commissioned officer should be able to send and read signalled messages, though not necessarily with the skill of trained signallers. Leaders must never think about their own units only; they should always consider the progress of the whole force of which their units form organic, integral, and constituent parts."

"NOTES ON MUSKETRY TRAINING"—(CHAPTER XII.)—"The higher part of the company training will be the field practices over varied ground and unknown ranges, in connection with some sound tactical idea, which should be varied for each practice. In this training all questions of combining firing with manœuvring, and of fire discipline, control, and direction, should be included, and it is especially important that this training should be constant and progressive, and be carried out continuously throughout the year, so that the men should never have time to forget the instructions imparted to them. On this subject the 'Musketry Regulations of the Native Army in India' very sensibly say":

Para. 107. Having been taught by drill and musketry instruction how to use his arms and shoot, and the formations in which to meet his enemy, the soldier has to acquire his training for battle or fire discipline.

The manœuvre-ground, and not the barrack-square and rifle-range, is the proper place to carry out the training for battle. Having been taught his drill on the parade-ground, and to use his rifle on the range, to aim correctly and shoot as well as he can be made to—instruction of supreme importance—he must be taught the combination of these on the open plains or the hillside, according to where he may happen to be stationed.

He must be taught to advance and retire when under fire, when to fire and how to fire, to move with a loaded rifle with safety to himself and his comrades, and be ready and able to inflict the greatest injury to his enemy. The allowance of ammunition made annually has to cover the soldier's preliminary range practice as well as his battle training, and the amount available for the latter is not always as much as could be desired. It must not be imagined that when this annual allowance is expended the soldier's battle training is done for the year.

In all schemes the *reason* for everything must be explained to the men by way of instructing them. The object and intentions of the commander in carrying out a certain manœuvre, why he halts, why he advances, why he opens and ceases fire, why he uses a certain description of fire—all must be carefully explained. . . .

"The only objections to field practices are (1) that the men never know whether their bullets have hit or missed the target aimed at; and (2) that the tactical part of the practice may be made to overshadow that of the musketry part of it. As regards the first objection, a man's excellence in shooting is tested at his range-firing; in field-firing he is made accustomed to war conditions of unknown ranges, uncertain objectives, and unknown effects. The second objection is met by proper superior supervision. It is the field and battle conditions of firing that we want to habituate our men and officers to, and if these are combined with simple, or with falling or collapsible targets, the effect of the shooting can be to some extent gauged, and the men made to feel the necessity for estimating correctly the range, for using the proper backsight elevations, and for observing as best they can the effects of their fire at all times. The highest results will be obtained if many of the field practices are carried out concurrently with the field training of the companies. **The novelty of field firing is apt to disconcert the men at first, but it very soon produces a marked difference in their individuality and an increased tendency to act more on their own initiative.* The curse of all our musketry hitherto has been its dulness and unreality, which tend to disgust the officers and men, and to make them callous about it. Unfortunately the value of fire-discipline, control, and direction, is not felt in peace manœuvres, and hence, in the past, it has been but very imperfectly practised. And therefore we want to do all we can to make musketry in all its branches as interesting to both the officers and men as possible, and thus to call out their individual co-operation in carrying it out properly and instructively."

"Coming now to the 'Schools of Musketry,' it is suggested that the Musketry Regulations, as such, and the way of making out forms and returns, etc., should not be taught there, but that much of what has been proposed for the regimental instruction should be gone over, theoretically and experimentally, *in far fuller detail and thoroughness, as well as the whole subject of collective firing.* In addition to this, the means of testing the sighting of rifles, and of finding the 'patterns' of a rifle and its ammunition, should be taught." (Note.—One generally associates the word "target" with a perpendicular object, the height of which is limited to that of a man.

A perpendicular target serves two purposes: (1) provides point to aim at, (2) indicates where some of the bullets fall. Why should not targets be horizontal on the ground, and used in conjunction with a small aiming target perpendicular in the centre of the former? The horizontal target should be the correct area to contain the "zone of effective fire" for each range used. For want of a better name, "Ground Carpet Target" could be used to describe the above method of obtaining "patterns" both of rifles and of machine guns. By the means of the "Ground Carpet Target" all kinds of problems in the application of Fire Tactics to Ground could be demonstrated and proved to officers, who would be able to quickly understand the foundation of fire tactics. The effectiveness of field firing could be increased by using the former target, because the reason for and the possibility of the closer grouping of shots would be demonstrated to the officers and men.—F. V. L.)

"The whole subject of the means of ensuring invisibility by means of natural and artificial cover, colour, background, avoidance of contrast of colours, distinctive marks, small size of target, avoidance of unnecessary movement when stationary and exposed, etc., should be fully discussed and illustrated. . . ."

"The effects of the shape of the ground on fire effects should also be illustrated. A horse or field artillery officer should be sent to every School of Musketry to give some lectures on the use and effects of artillery fire. . . ."

"As regards the utilization of the amount of ball ammunition allowed to each man annually, a very few rounds a year are sufficient for use *by trained soldiers* over measured ranges. Every round possible should be kept for field practices against small and rapidly moving and disappearing targets (why not Ground Carpet Targets at unknown ranges?—F. V. L.), placed either in the open or behind cover, over known ranges on varied ground, and be used in connection with a variety of tactical exercises (*e.g.*, attacking, defending, repulsing, night-fighting, pursuing, retreating by alternate units, etc.)—all of which, however, need not be attempted with ball ammunition in each year, but some in one year and others in another. The allotment of only a limited quantity of ammunition for such practices gives the troops a very false impression of the amount and intensity of the fire that would probably be required on active service to overcome a well-armed and well-trained enemy."

"All the men taken out on any one day need not be made to fire on that day; by making different men fire on different days, with the remainder looking on or firing blank, we can extend the tactical exercise and their instructional influences over a considerable period of time; * *but care should be taken to make these exercises progressive—that is, rising by degrees from the simpler to the more complex ones.* We must remember that, important as individual marksmanship is, yet it is not everything. *How to employ collective fire*, at what distance fire should be opened when attacking and defending under different circumstances, how to avoid losses while inflicting them on the enemy, how to make use of ground and cover while moving forward, and so on, are all matters of equal importance, and contribute as much as, if not more than, individual marksmanship to the winning of battles." (NOTE.—Logically, the fire of machine guns should, in every case, be substituted for concentrated fire of infantry.—F. V. L.)

* "*The art of fighting battles is a complex one, being composed of many combining factors, each of which must be given its due weight and proportion, and no more nor less; if this is not done, then the danger is at once incurred of introducing a 'topsidedness' in the resulting mode of conducting the fight, with a corresponding loss of effectiveness.* It is of vital importance to obtain and maintain order and control in the firing-line, to direct and control its fire, to ensure the best tactical use of it, to ensure the men moving forward when required, and in the right direction, to see that they make the best possible use of all cover, to maintain the necessary coolness and judgment for obtaining an effective fire, and to get the men of the various companies and battalions engaged to *work together* for the common end."

"In such exercises those performed as company exercises are of the highest importance, battalion and brigade exercises being only of use in training the smaller units to work together in organic union. (NOTE.—The above is applicable to a company of the Machine-Gun Corps, which has to work in organic union with the four battalions of a brigade.—F. V. L.) And in these higher exercises it is of the utmost importance that the proper value be given to artillery fire. **It is absurd to assume that the proper artillery preparation had been completed in a few minutes in order to let the men get back to their dinners.*"

"We now come to our last, but very practical, point, and that is the provision of the ground necessary for the proper training of our troops. Assuming that all the coming (*circa* 1903) reforms in organization, armament, equipment, drills, tactics, methods of training, etc., are the very best that can be devised, yet of what real advantage will they prove if our troops are not allowed to practise them annually on a wide and unrestricted scale?"

"If the coming reforms, or even if the Musketry Regulations, are to result in any real benefit to our Empire, the military authorities must be given a free hand, in certain defined areas if need be, to practise their profession over sufficiently large areas of ground, under as near an approach to war conditions as possible, and without any restrictions of forbidden areas. Our Army has never yet had such opportunities for preparing itself for the very purpose of its existence, and until our troops can freely operate on certain areas of sufficient size and suitable character, and be allowed within that area to manœuvre, shoot, and dig as may be required, we shall enter into the next war as unprepared and as untrained as we did into the Boer War."

(NOTE.—The late Colonel Mayne was absolutely correct in the above sentence. *Neither the machine gun nor its officers have ever had adequate facilities of ground in Great Britain, to work out the highest fire tactics of that weapon in large enough numbers.* Such a firing ground should be at least four miles square (the machine gun fires up to

3,000 yards) with quarters for the troops *outside* the area. The maximum injury to the enemy cannot be obtained from any weapon without many experiments with unlimited conditions; and in the case of machine guns these would be (1) use of gun at all ranges up to the extreme one and on all shapes of ground, (2) the use of unlimited ammunition, (3) the use of an unlimited number of machine guns. The personnel for these experiments would in the first case be drawn from officers, non-commissioned officers, and men of two years' service in machine-gun sections. In course of time the lieutenants would grow into captains, majors, colonels, and generals, while the non-commissioned officers would be promoted to Warrant rank. As in the case of Sir John Moore's Light Infantry, the sergeants must be young, clever, and active men ready for progress in all directions. To return to the tactical firing ground four miles square, the setting apart of four such areas in Great Britain and one in Ireland would ensure that our Fire Tactics always had a chance to progress and to get the most effective injury on our next enemy. Our Fleets and Squadrons have the ocean to work out manœuvres on, so why cramp our Soldiers? The above five areas, with barracks *outside* each of them, would be a very cheap way to save hundreds of thousands of lives in the future.—F. V. L.).

The Notes on German Musketry Training are so much to the point, that no apology is needed in giving extracts. Because we do not approve the German method of carrying out the training of their recruits, that is no reason why we should not learn from them all we can about Training Grounds and their Works.

"After the adoption of the Mauser rifle, and the advances made in the musketry instruction of the infantry, and in the study of the effects of rifle fire, the necessity for practising troops in the execution of field firing (battle-firing) under conditions approaching as nearly as possible to those of war, was recognized. To give this instruction, the Germans at first made use of the firing grounds that their artillery already possessed, and, beginning in 1883, very large sums were annually voted by their Parliament both to augment the somewhat limited dimensions of these firing grounds and to establish new ones. It was in this way that the firing grounds at Haguenau, Lockstadt, Zeithain, Griesheim (Darmstadt), Jüterbog, Tegel, Falkenberg, and Wahn were greatly extended, and those at Gruppe, Hamerstein, and Altenberg were brought into existence. . . . Such were the motives that have induced the German War Office since 1891 to establish a number of sufficiently extensive manœuvre grounds, where field firing and manœuvres over varied ground could be undertaken by all arms, either separately or combined, at all seasons of the year, and without having to consider agricultural interests. Chosen as much as possible in sparsely populated regions, and where the soil is of relatively small value, these manœuvre grounds ought, in order to fulfil their object, (1) to be situated in more or less central positions as regards the garrisons who have to make use of them; (2) to permit of the fire of infantry and artillery in several directions; (3) to have sufficiently large dimensions to enable large units composed of all three arms to be massed and manœuvred on them; and (4) to present a sufficient variety in the accidents of the ground to represent the ordinary conditions of war to the troops, and that without preventing the movements and employment of cavalry."

"Their theoretical extent was fixed at 21·8 square miles, or about 4·67 miles a side, but at the outset only one manœuvre ground, that of Lockstadt, approximately reached those dimensions. Thus, in 1896-97, Germany possessed fourteen manœuvre grounds, and two garrison artillery firing grounds. The sums voted from 1891-92 to 1895-96 for these purposes amounted to £2,410,480, and the budget for 1896-97 provided for a further grant of £446,950; and an additional sum of £3,507,430 was needed to make a complete provision for all the German (army) corps."

"The number and extent of the firing and manœuvre grounds, as well as the large sums that are voted yearly for them, give an idea of the importance that the Germans attach to field firing, and that they should be used for this purpose at all seasons of every year for the manœuvre of regiments, brigades, and divisions, as well as of forces of all three arms —it being rightly considered that in this way they will obtain the best possible school for the tactical instruction of their officers and the manœuvring aptitude of their troops. Even a very slight study of the German Infantry Regulations will show us what a constant, and even absorbing, preoccupation the Germans have for securing the preponderance in action of rifle fire, combined with a violent offensive forward movement. Theoretically these tactical proceedings and methods of instruction are almost solely concerned with the view of obtaining the superiority of the fire, which the Germans consider to be the chief guarantee of success. Practically, we find the same thing in the scrupulous care they devote to musketry instruction in order to get the best results possible from their fire; in the flexibility of their formations, and the rapidity with which they are taken up, by means of which they hope to be able to surprise the enemy; and in the immense firing grounds and extensive manœuvre areas of Arys, Jüterbog, Elsenborn, Döberitz, etc., which are veritable schools for war, and for the provision of which they have not recoiled from any sacrifice. They consider that success belongs beforehand to those who possess the best musketry instruction, the most severe fire-discipline, and the best-managed direction of fire."

"We may not agree with the dense firing lines made use of in the German Army and the still denser formations laid down for their support, after our South African experience—and, indeed, the Germans do not altogether adhere to them—but we cannot but help admiring the thoroughness and consistency with which they put their guiding principles into practice. * *Furthermore, there is much in their instructional* and tactical *methods that we can copy with advantage.* But in doing this we must be careful not to copy the mere outward forms of German practice, as we have so often done, *without bearing in mind the preparation they have made for the application of these forms.* The high cost of land in Great Britain, *the idea that shooting game is of more importance than giving* *the best training possible to our men to uphold their country's honour and policy*, the grudging way in which money is doled out in peace time for military training for fear of party political capital being made out of it, etc., are so many various factors that have prevented our Army having either the use, even on payment, of the country for manœuvring over in large masses once a year for field firing on a suitable scale; or the proper provision of ordinary ranges near at hand to our barracks. The result is that, as the bulk of our troops cannot get *daily* access to convenient ranges, the annual musketry practices for each company are got through in a few days, in any weather, while for the rest of the year no ball-firing takes place, unless some lucky general can manage to get one day's field firing in some country out of Great Britain. What is often practical under the name of 'field firing' is an utter misnomer, because true 'field' conditions are absent. But in India and in South Africa, where a very considerable portion of our troops will probably be stationed in the future, for political and strategical reasons, we have very great possibilities for the proper training of our troops, and all that is needed is the purchase of sufficiently large areas of suitable land, while in Great Britain a 'Manœuvre Act' is urgently needed, giving the use of a wide tract of country, once a year, for the manœuvring of our troops at home in large masses under war conditions."

"But in spite of all the disadvantages under which our Army has to be trained, our small company organization, combined with the zeal and energy of our officers, have always enabled our troops to give a good account of themselves in war, though at the cost of a voluntary noble sacrifice of life that might have been rendered unnecessary under better conditions; indeed, *our Army has a high amount of efficiency out of all proportion to the small advantages offered to them for this purpose.* No doubt our national love of games, and especially of games that need endurance and courage, and that have an element of danger in them, and which train men to 'play together,' has given the necessary basis on which our officers have been able to build so successfully. But the very conditions and limitations of our training render it necessary for us to be most careful in blindly following the German principle of '*free*' tactical procedure, as so many hold up as the one ideal of battle. In Germany it has been preached by pen and tongue, but in practice, with the thorough and continuous practice that the German officers and men undergo, the resulting tactical procedure is not so very free as one is apt to consider; *they have, in fact, become more or less cast in definite moulds* from the fact that their troops always serve together and are always under the same leaders in peace and war. But even with all these advantages the inherent dangers of a free tactical procedure have so strongly shown themselves in the German annual manœuvres, that, in Germany itself, its advocates are fast losing ground. The Germans have adopted the principle of 'organized battlefields,' and this is rapidly leading them back again to a definite and recognized normal initial battle organization for all units, capable, however, of being modified to suit any special and abnormal local circumstances. And if this necessity for a return to the older ideas is being found necessary for the German Army, with all its enormous advantages for training, and which is daily gaining ground rapidly among them, judging by the military papers of the day, much more necessary is this necessity for our own Army, with its limited advantages for proper tactical and strategical training."

INDEX

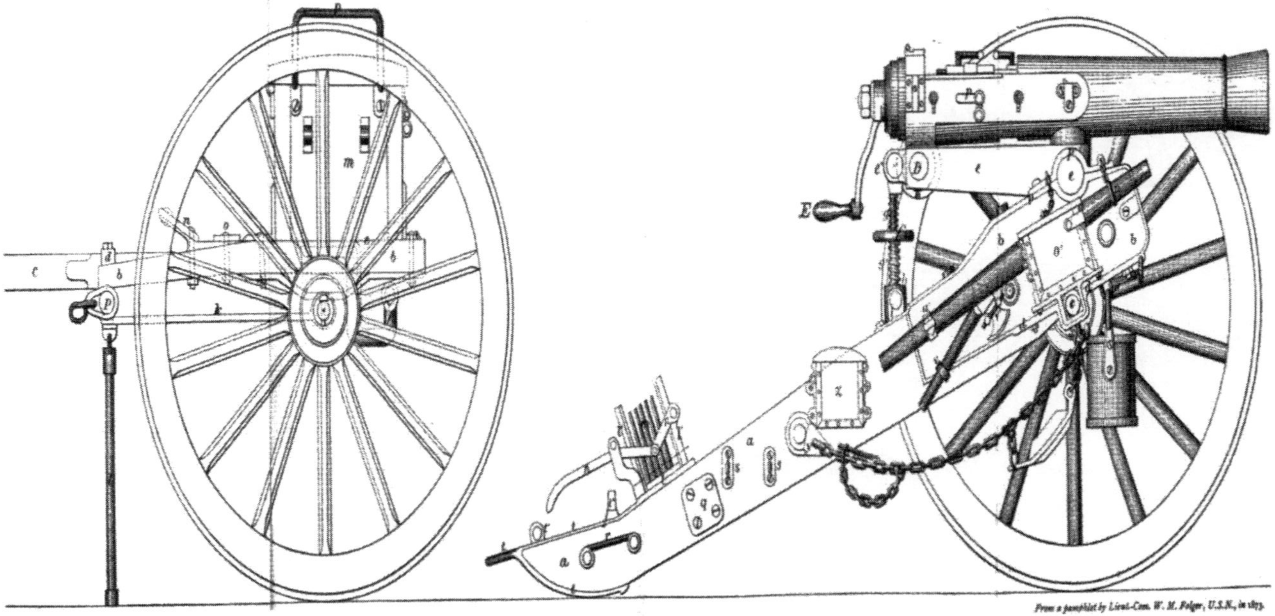

From a pamphlet by Lieut.-Com. W. M. Folger, U.S.N., in 1873.

The French Mitrailleuse (or Canon à Balles) of 1870, showing Trail Mounting and Limber.

There are twenty-five breech-loading barrels of 950 millimetres in length and 13 millimetres in calibre. Those parts marked *l*, *m*, *n*, compose the extractor for clearing the empty cases from the cartridge magazine.

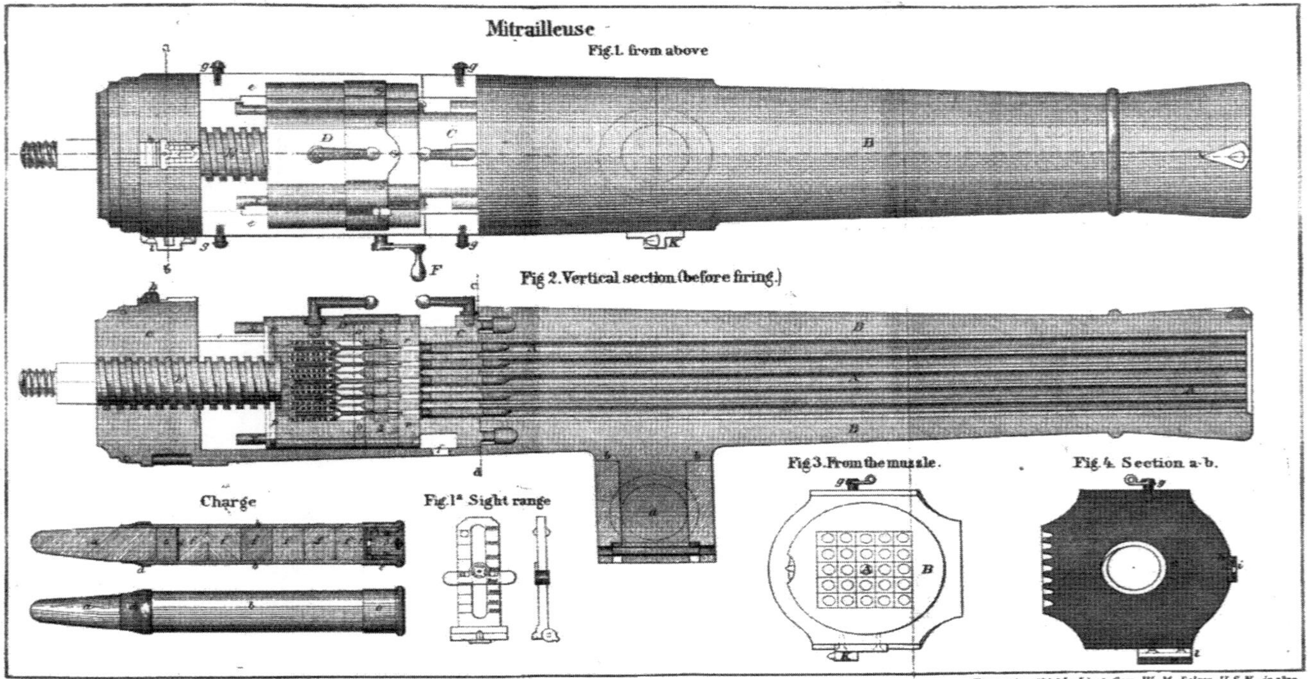

French Mitrailleuse of 1870; Calibre 13 Millimetres.

With well-trained gunners at least 100 rounds could be fired in one minute. Sighting is up to 1,300 yards.

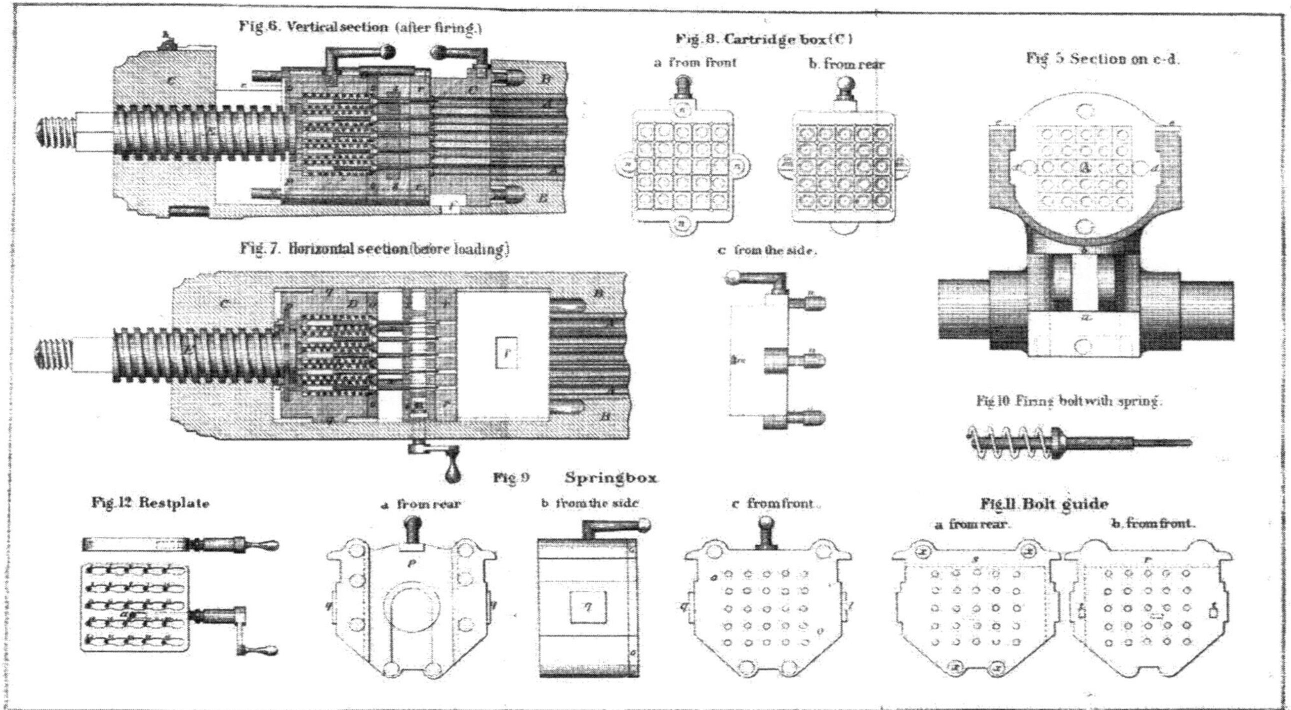

From a pamphlet by Lieut.-Com. W. M. Folger, U.S.N., in 1873.

The Mechanism of the Mitrailleuse of 1870; Calibre 13 Millimetres.

In modern nomenclature the above figures would be called: (8) cartridge magazine, (12) trigger plate, (10) striker, (11) bolt head.

4

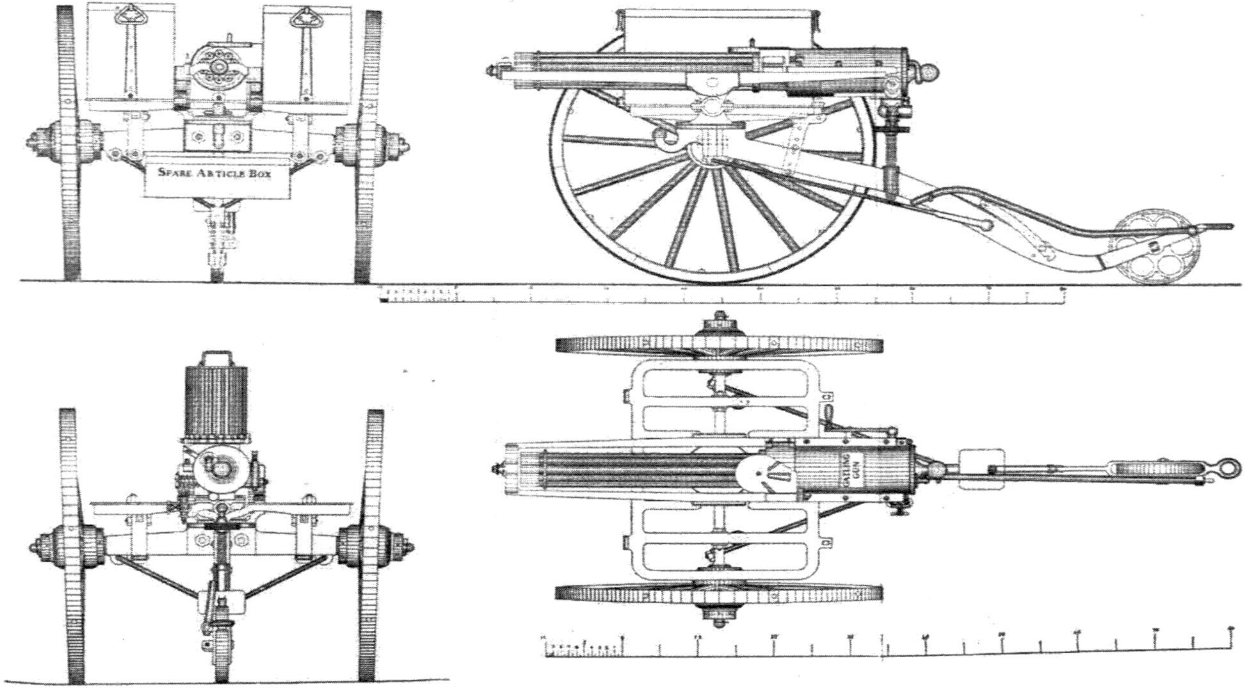

United States Naval Landing Carriage for Long Gatling Gun. Per U.S. Navy Bureau of Ordnance, 1875.
Feed by gravity drum (each 400 rounds), two to each gun. Same ammunition as Remington rifle: ·50-inch.

From "Nordenfelt on Machine Guns," 1884.

The 'Alt Magazine Carriage, *circa* 1884, One of the Two Nordenfelt Rifle-Calibre Machine Guns belonging to the Machine-Gun Section of the Central London Rangers (Volunteers).

The gun had five barrels and the carriage held 5,000 rounds of ·45-inch ammunition.

From "Nordenfelt on Machine Guns," 1884.

The Machine-Gun Section, with Nordenfelt Machine Guns on Alt Magazine Carriages, of the Central London Rangers forming "Rallying Squares"; *circa* 1884.

J.U.S.I.

Gardner Machine Gun, Five Barrels, Gravity Feed, Firing Handle on Right Side.

This is a very early form of the Gardner system ; *circa* 1885.

Military Hotchkiss Revolving Field Gun: Five Barrels; Calibre, 37 Millimetres (1·45 Inches), which together can fire Eighty Shells per Minute to over Three and a Half Miles with Accuracy.

By the Hotchkiss Co. at St. Denis, Paris, *circa* 1885. (From *Engineering*, January 23, 1891.)

Hotchkiss Revolving Gun on Ship Mounting; Calibre, 53 Millimetres, or 2·1 Inches; Five Barrels.

By the Hotchkiss Co. at St. Denis, Paris, *circa* 1885. (From *Engineering*, January 23, 1891.)

FIG. 1.

FIG. 2.

From "Scientific American Supplement."

Maxim A.R.C.M.Gun on an Ingenious Form of Steel Field Carriage.

Fig. 1 shows it arranged for horse draft, Fig. 2 for man draft; *circa* 1887.

Fig. 3.

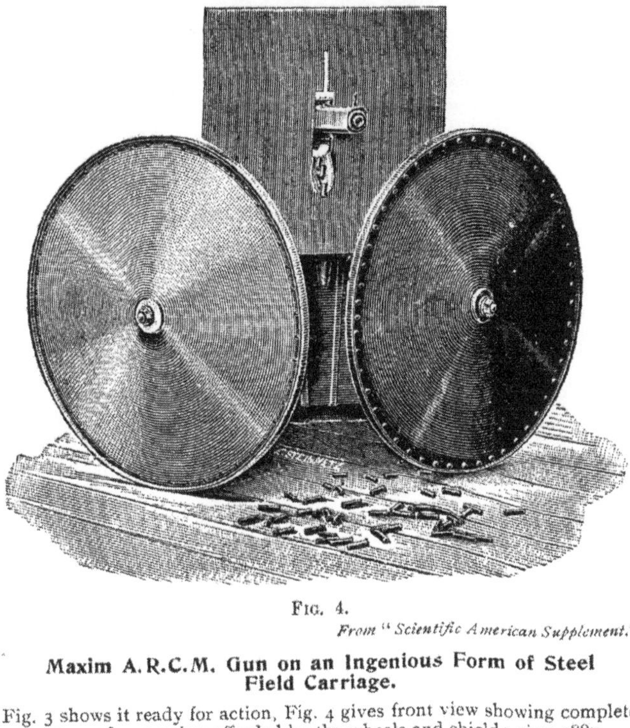

Fig. 4.

From "Scientific American Supplement."

Maxim A.R.C.M. Gun on an Ingenious Form of Steel Field Carriage.

Fig. 3 shows it ready for action, Fig. 4 gives front view showing complete nature of protection afforded by the wheels and shield; *circa* 1887.

Photograph by Gregory and Seeley, Ltd., 51, Strand, London.

Machine Guns at School of Musketry, Hythe.

From left to right: Maxim; three-barrel Nordenfelt on infantry field carriage, Mark I.; Maxim on armoured tripod; Maxim on infantry carriage, Mark I.; two-barrel improved Gardner on parapet mounting; *circa* 1889.

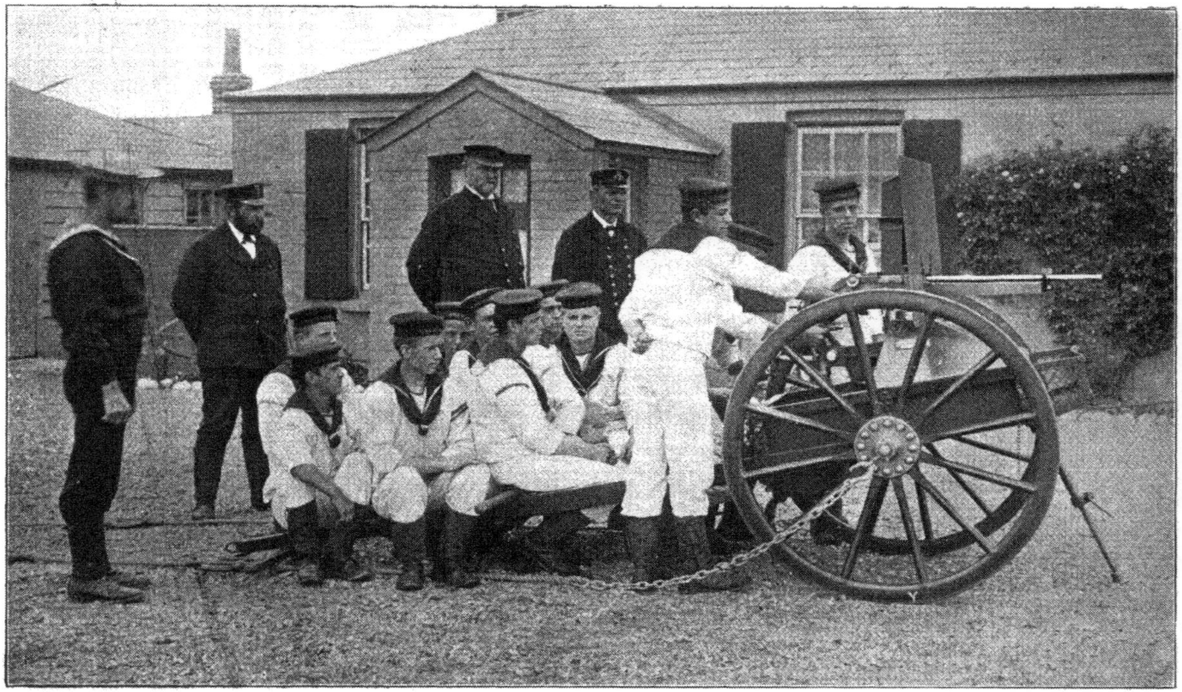

Symonds and Co., Portsmouth.

Land Service Carriage (issued about 1888). Five-Barrel Nordenfelt Machine Gun with Hopper, Shield, and Prop-Stick shipped.

The gun's crew (H.M.S. *St. Vincent*, at Portsmouth) were following out the old drill in taking shelter behind the carriage.

Photograph by Gregory and Seeley, Ltd., 51, Strand, London.

Maxim R.C.A.M.Gun on Cavalry Machine-Gun Field Carriage, ·303-Inch, Mark I. (*circa* 1889), with the *Addition* of a Steel Shield all over Back of the Carriage. (Probably 1st Dragoons.)

Photograph by Gregory and Seeley, Ltd., 51, Strand, London.

Two Five-Barrel Nordenfelts on Landing Carriages in Square to receive Dervishes.

The first pattern of carriage (1887) is in the centre, the later pattern is on the right of the square. This illustrates the effect on fire tactics of the Gordon Relief Expedition in 1884-85, and was taken at Whale Island, Portsmouth, *circa* 1890.

Photograph by Gregory and Seeley, Ltd., 51, Strand, London.

Nordenfelt Machine Gun, Three Barrels, on Infantry Field Carriage, Mark I. (November 30, 1888).

Training Battery of Royal Marine Light Infantry at the Plymouth Barracks.

From left to right: Four-barrel 1-inch Nordenfelt; 3-pounder Hotchkiss Q.F.; five-barrel ·45-inch Nordenfelt; ·45-inch Maxim on cone mounting; *circa* 1892.

Photograph by Gregory and Seeley, Ltd., 51, Strand, London.

Six-Barrel (Improved) Gatling Machine Gun on Infantry Carriage, belonging to the 3rd Battalion London Rifle Volunteers; *circa* 1895.

This gun has the Accles Feed System.

Photograph by Maxim-Nordenfelt.

A Light Pattern Maxim Gun, weighing 25 Pounds, and Tripod weighing 15 Pounds, carried together in a Back Pack.

Note the barrel has no water-jacket. This was brought out about 1895 by the Maxim-Nordenfelt Guns and Ammunition Co., Ltd., of London.

Photograph by Maxim-Nordenfelt.

A Light Pattern Maxim Gun, weighing 25 Pounds, Tripod weighing 15 Pounds, the Whole of which can either be carried by One Man on his Back or on a Horse in Addition to the Rider.

In the latter case the gun and tripod are carried on the off-side, and the ammunition on the near side, both in perpendicular leather boots. This was brought out about 1895 by the Maxim-Nordenfelt Guns and Ammunition Co., Ltd., of London.

Photograph by Gregory and Seeley, Ltd., 51, Strand, London.

Carriage, Field, Machine Gun, Cavalry, ·303 Maxim, Mark I.; *circa* 1895.

The Machine-Gun Section, 6th Dragoon Guards, at standing drill.

Photograph by Gregory and Seeley, Ltd., 51, Strand, London.

Photograph by Gregory and Seeley, Ltd., 51, Strand, London.

Two Light Air-Cooled Maxim Automatic Machine Guns, 1896, at the Naval and Military Tournament, London.

Mounted on a tricycle made for two, in such a way that the guns can be fired from the machine or separated and mounted on tripods. The gunners belong to the 2nd Battalion Tower Hamlets Volunteers, while the escort belong to the 26th Battalion Middlesex Volunteers (Cyclists).

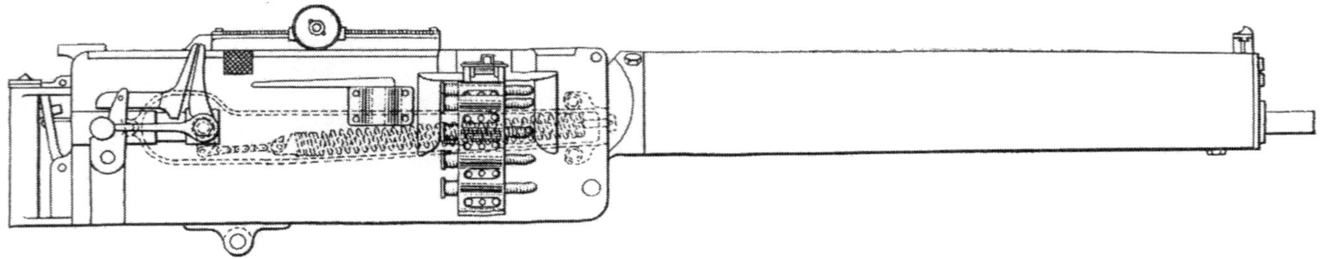

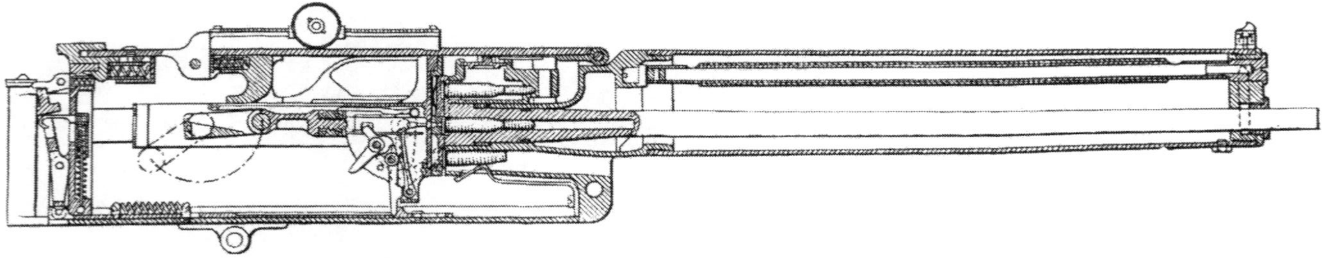

Maxim-Nordenfelt Guns and Ammunition Co., 1897.

Longitudinal Elevation and Section of ·45-Inch Maxim R.C.A.M.Gun ; Weight, 54 Pounds.

An early pattern issued to the British Service ; guns of this pattern are still to be found in many parts of the Empire.

Photograph by Colt Patent Firearms Manufacturing Co., U.S.A.

Light Gatling Gun, ·300 Inch, on Trail Mounting.

The limber carries 9,840 rounds, gravity straight feed-case holding 40 rounds. Made by Colt's Patent Firearms Manufacturing Co., Hartford, Conn., U.S.A. The pattern used by Lieutenant J. H. Parker in 1898 at Santiago.

Photograph by Colt Patent Firearms Manufacturing Co., U.S.A.

Light Gatling Gun, ·300 Inch: Pattern used at Santiago in 1898.

Lord Dundonald's Machine-Gun Galloping Carriage, with Trail connected to the Horse; *circa* 1898.

The gun is the light air-cooled Maxim with pistol grip. (From *The Engineer*.)

Photograph by Gregory and Sieley, Ltd., 51, Strand, London.

Fig. I.

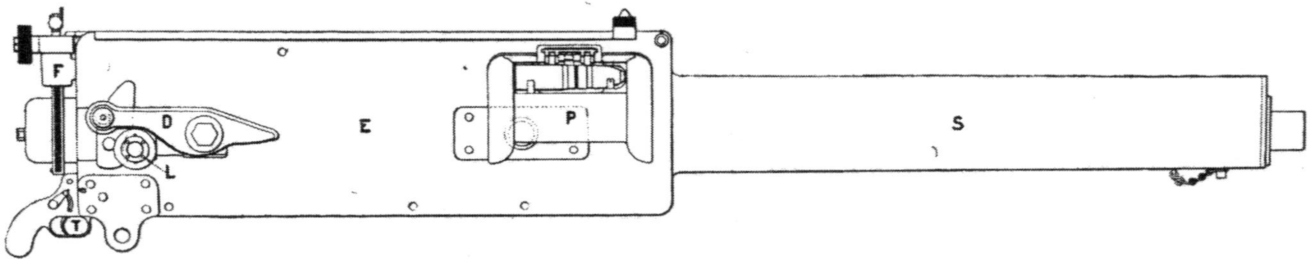

Fig. II.

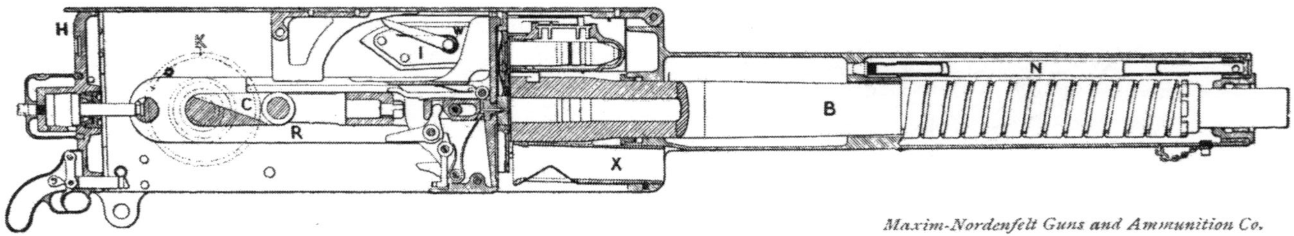

Maxim-Nordenfelt Guns and Ammunition Co.

Maxim 37-Millimetre Automatic Gun (1·457 Inches; Weight of Shell, 1 Pound), 1898.

Weight of gun 3 cwt. 2 quarters 24 lb. This is the "Pom-Pom" of the South African War, 1900. Officially known as 1 Pr. Q.F. Maxim guns of 1 inch and 1·5 inch were first made about 1886.

Maxim-Nordenfelt Guns and Ammunition Co.

Trail Mounting and Limber for Maxim (1·457-Inch) Automatic Machine Gun, the latter carrying 300 Rounds.

Designed and made by the Maxim-Nordenfelt Guns and Ammunition Co., *circa* 1898. The Boers used many of these guns in 1900.

Maxim-Nordenfelt Guns and Ammunition Co.

Maxim 37-Millimetre Automatic Machine Gun, on Cone Mounting with Shoulder-Piece, 1898.

Each belt holds twenty-five rounds of 1·457-inch. This gun can also be provided with a field trail and limber. (See Plate 80.)

Photograph by Gale and Polden.

Mark III. Maxim Machine-Gun Infantry Carriage, with Men of the Section and a Draft Mule.

This Mark was first issued in 1900, and about 1912 those units still having this carriage, had a Mark IV. tripod issued for each,

Early Heavy Form of German Maxim R.C.A.M.Gun and Sledge Mounting, first issued about 1900.

Front legs adjusted for firing sitting. Note the water-jacket is connected to mounting by gimbals.

Early Form of German Maxim R.C.A.M.Gun and Sledge Mounting, first issued about 1900.

Early Heavy Form of German Maxim R.C.A.M.Gun and Sledge Mounting, first issued about 1900.
Front legs adjusted for dragging along the ground.

Early Heavy Form of German *Maxim* R.C.A.M.Gun and Sledge *Mounting*, first issued about 1900.

Early Heavy Form of German *Sledge* Mounting, with Wheels *shipped*, carried on a Pack-Horse ; *circa* 1900.

This is a top pack, instead of a side pack, the latter being used by most armies,

Photograph by Gregory and Seeley, Ltd., 51, Strand, London.

British Naval Maxim R.C.A.M.Gun on Mark III. Tripod (October 26, 1901), at Whale Island.

Note the gun and tripod are transported by two seamen with a carrying pole, so that both the former are ready for action without

Bergman's System of R.C.A.M.Guns, Model 1902.

The cover is open to show mechanism ; the tripod legs are not adjustable ; there are traversing stops.

Bergman's System of R.C.A.M.Guns, Model 1902.

The feed is a metallic belt ; note shape of belt-box.

Swiss Machine-Gun Mounted Company (1902), showing Pack-Horse carrying Maxim R.C.A.M.Gun on Side of Saddle.

The leading-stick is for use when the driver is mounted.

... Pack Horse carrying Fabric Feed-Belts in Eight Boxes

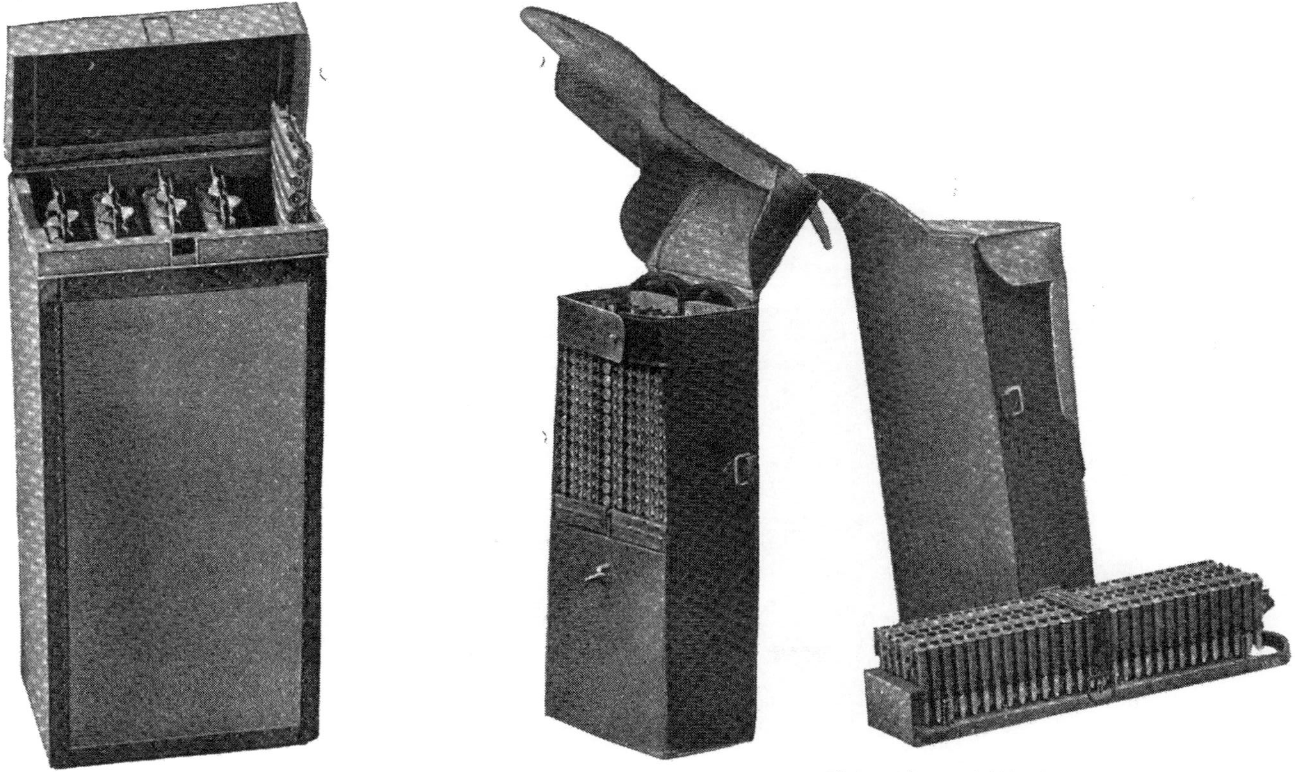

Photograph from the Hotchkiss Company of Saint-Denis.

Hotchkiss Feed-Strip Boxes (1903 and 1908).

Wooden box for infantry machine-gun sections, holding ten strips of thirty rounds each, total 300; one pack-horse carries six boxes, total 1,800 rounds. Leather holster carriers for use with cavalry.

Photograph from "Engineering," February 20, 1903.

Hotchkiss Automatic Machine Gun.

Sometimes referred to as the Odkolek gun. The usual system of loading is by feed strips containing thirty rounds, but above is shown a steel feed belt carrying 250 steel cartridge-holders, which are hinged together. Introduced in the French Army in 1899.

Photograph from "Engineering," February 20, 1903.

Hotchkiss Automatic Machine Gun.

On an early form of armoured car. Note tripod strapped on the side.

Photograph from "Engineering," February 20, 1903.

Hotchkiss (One-Pounder) 37-Millimetre Automatic Machine Gun.

Naval mounting.

Photograph from "Engineering," February 20, 1903.

Hotchkiss One-Pounder 37-Millimetre Automatic Machine Gun.

Field mounting ; ammunition in clips of eight rounds ; breech closed.

Photograph from "The Illustrated London News" of October 16, 1915.

Madsen R.C.A.M.Gun, Recoil Operated.

First introduced into England by the Rexer Arms Co., in 1904, when it competed at Bisley. Weight, 17½ pounds. Invented by Mr. Schouboe, a Danish engineer, and by him named Madsen, after the then Danish War Minister. Known in the U.S.A. as the "Benet-Mercier." The above gun was photographed by the Germans, after being captured from the Russian cavalry.

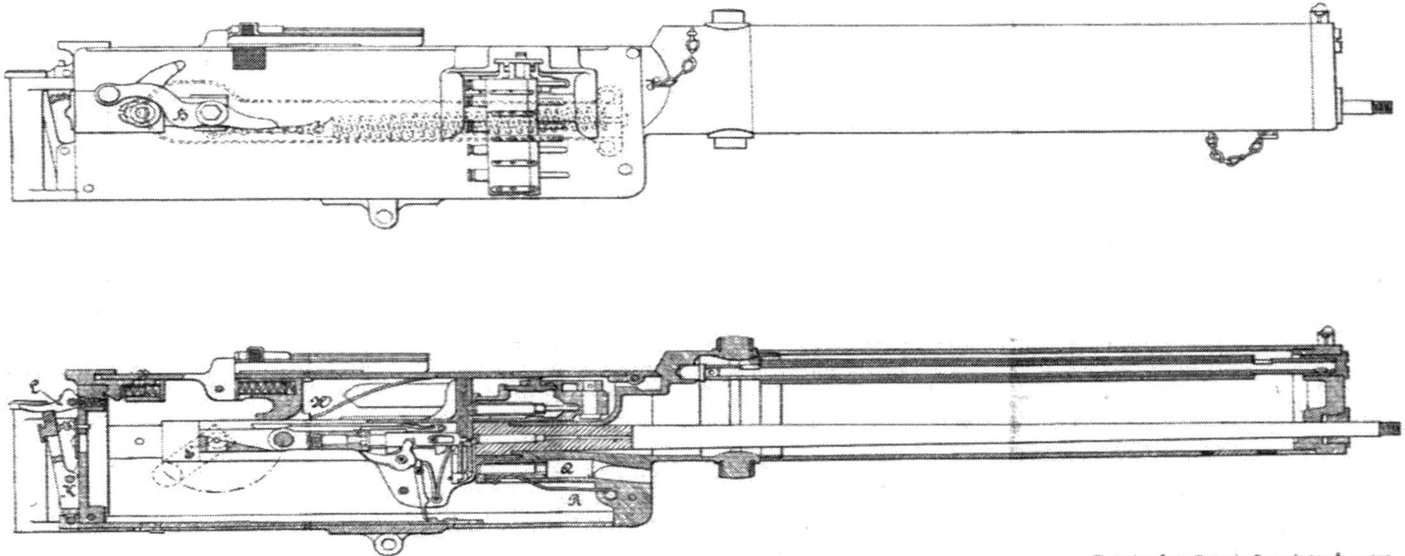

Drawing from Captain Braun's work of 1905.

Longitudinal Elevation and Section of German Maxim R.C.A.M.Gun.

There are two trunnions on the water-jacket instead of a crosshead pinhole, to connect gun to mounting. These two form part of the gimbal.

Fig. 1.

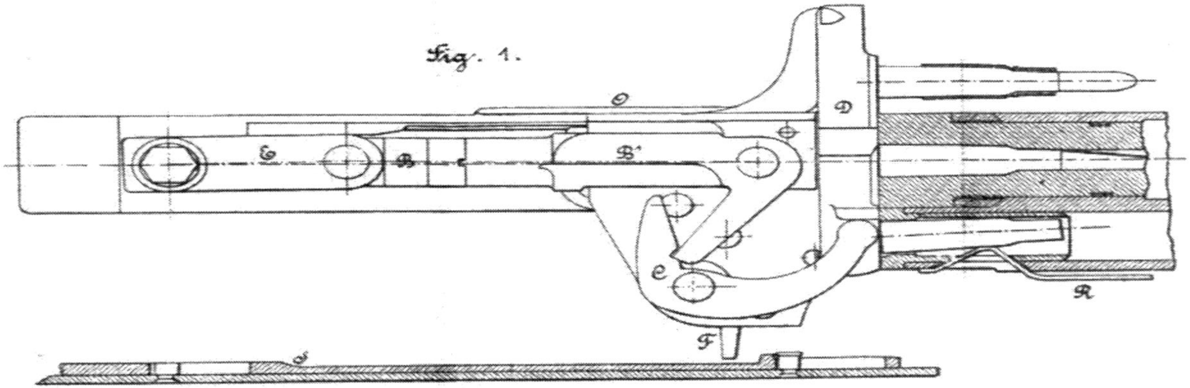

Fig. 2.

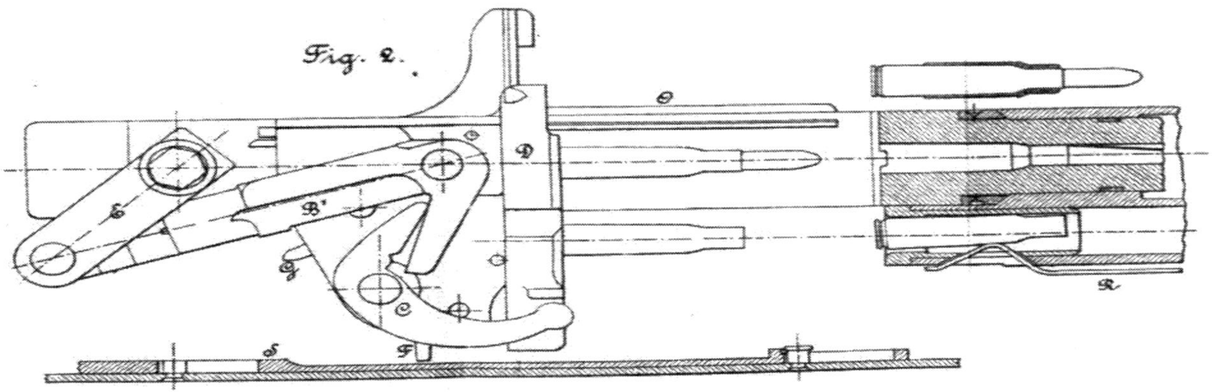

Section of Lock, Closed and Open, of German Maxim R.C.A.M.Gun.

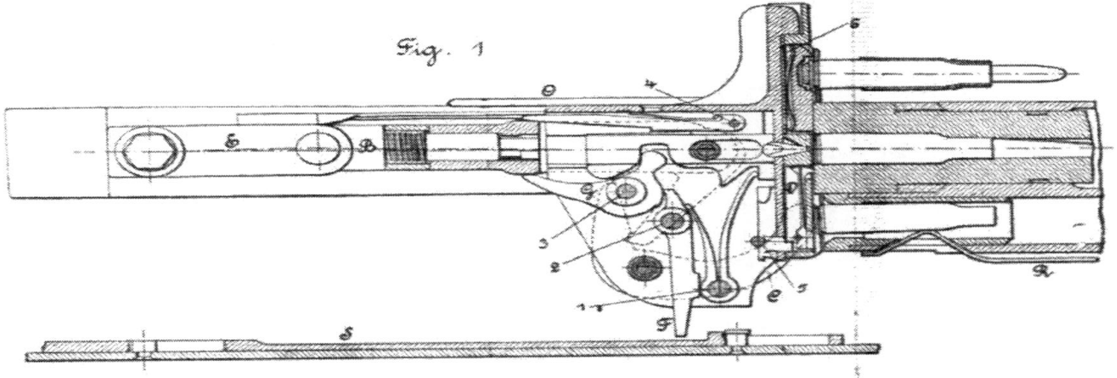

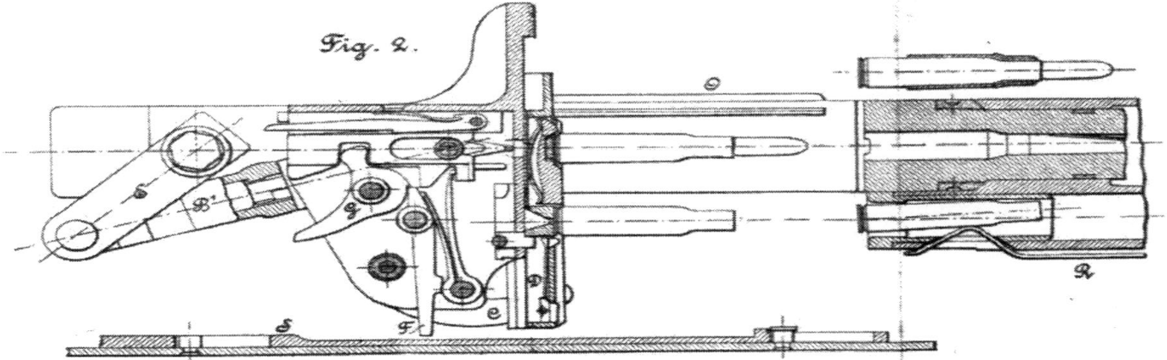

Drawing from Captain Braun's work of 1905.

Elevation of Lock, Open and Closed, of German Maxim R.C.A.M.Gun.

The base of the cartridge-case is grooved and flush instead of having a projecting base.

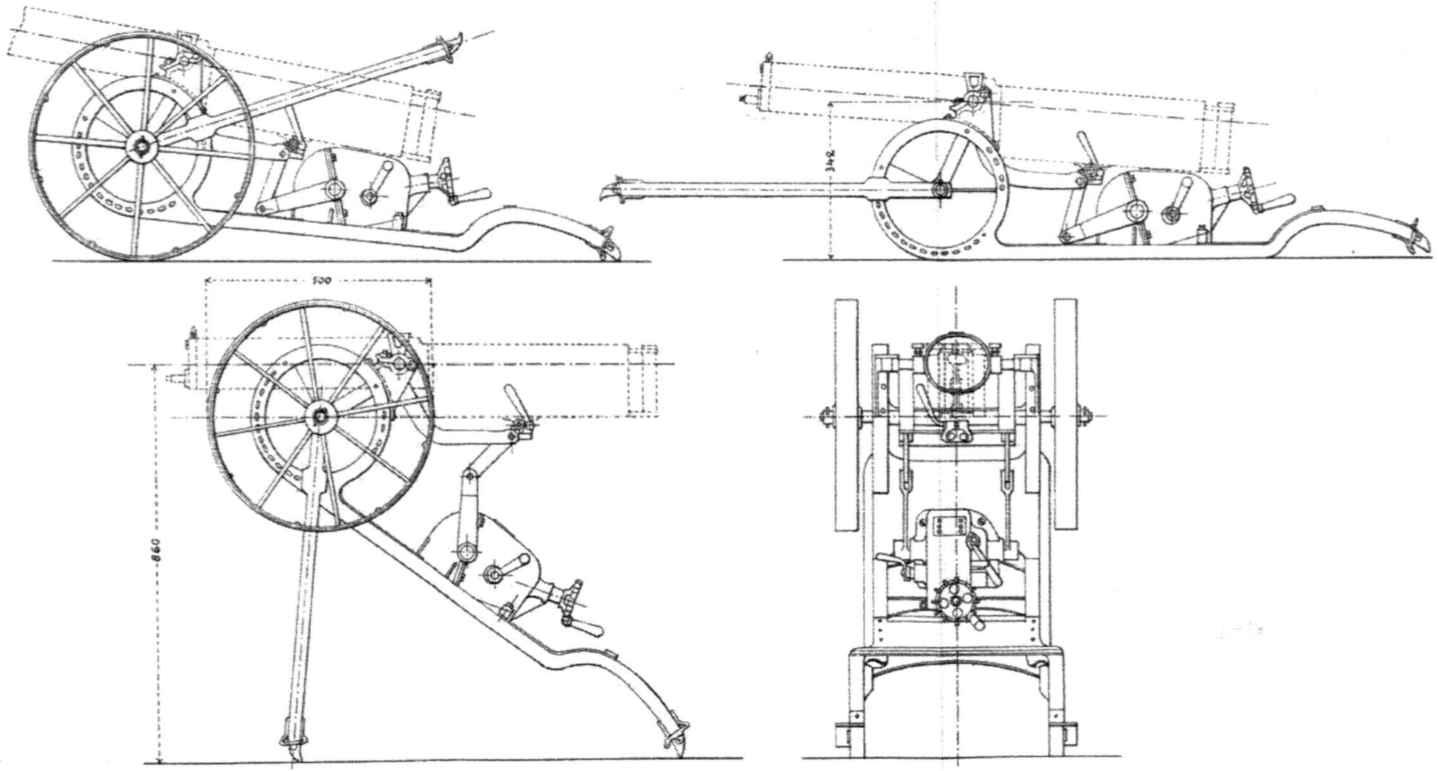

Improved Pattern of Sledge Mounting for German Maxim R.C.A.M.Gun, which can be used with or without Wheels.

Three elevations from the side and one from the rear. Note improved elevating gear and traversing clamp.

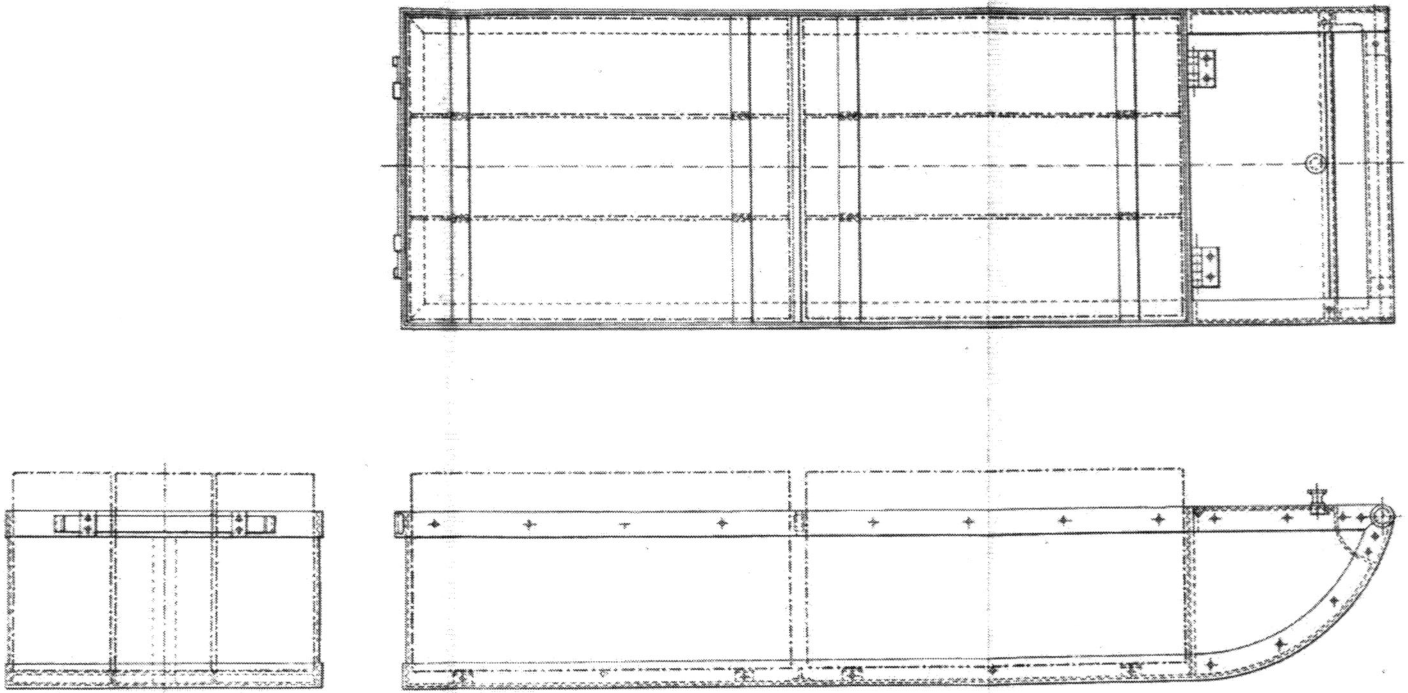

Steel Ammunition Sledge to carry Six Fabric Belt-Boxes for German Maxim R.C.A.M.Gun.

By this means one man can transport six boxes instead of two only. Plan, side elevation, and rear elevation.

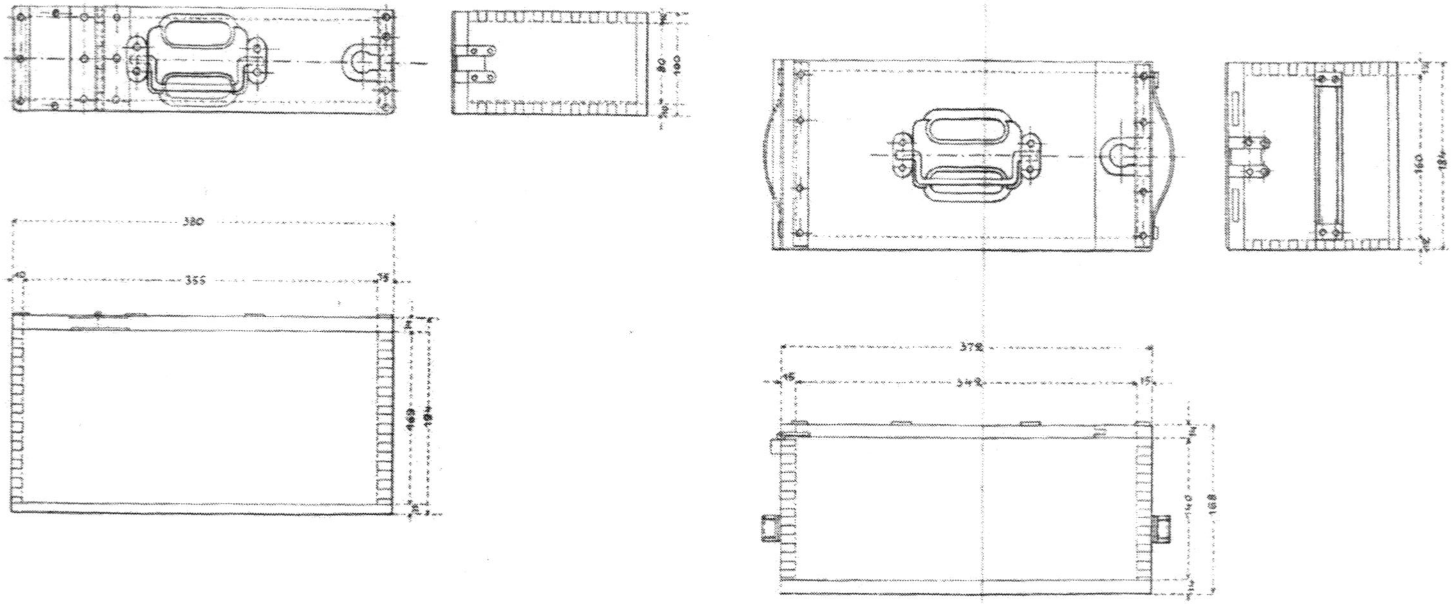

Wooden Box for Fabric Feed-Belt, Tool or Store Box for German Maxim R.C.A.M.Gun.

Note the handles fold flush with the top of lid.

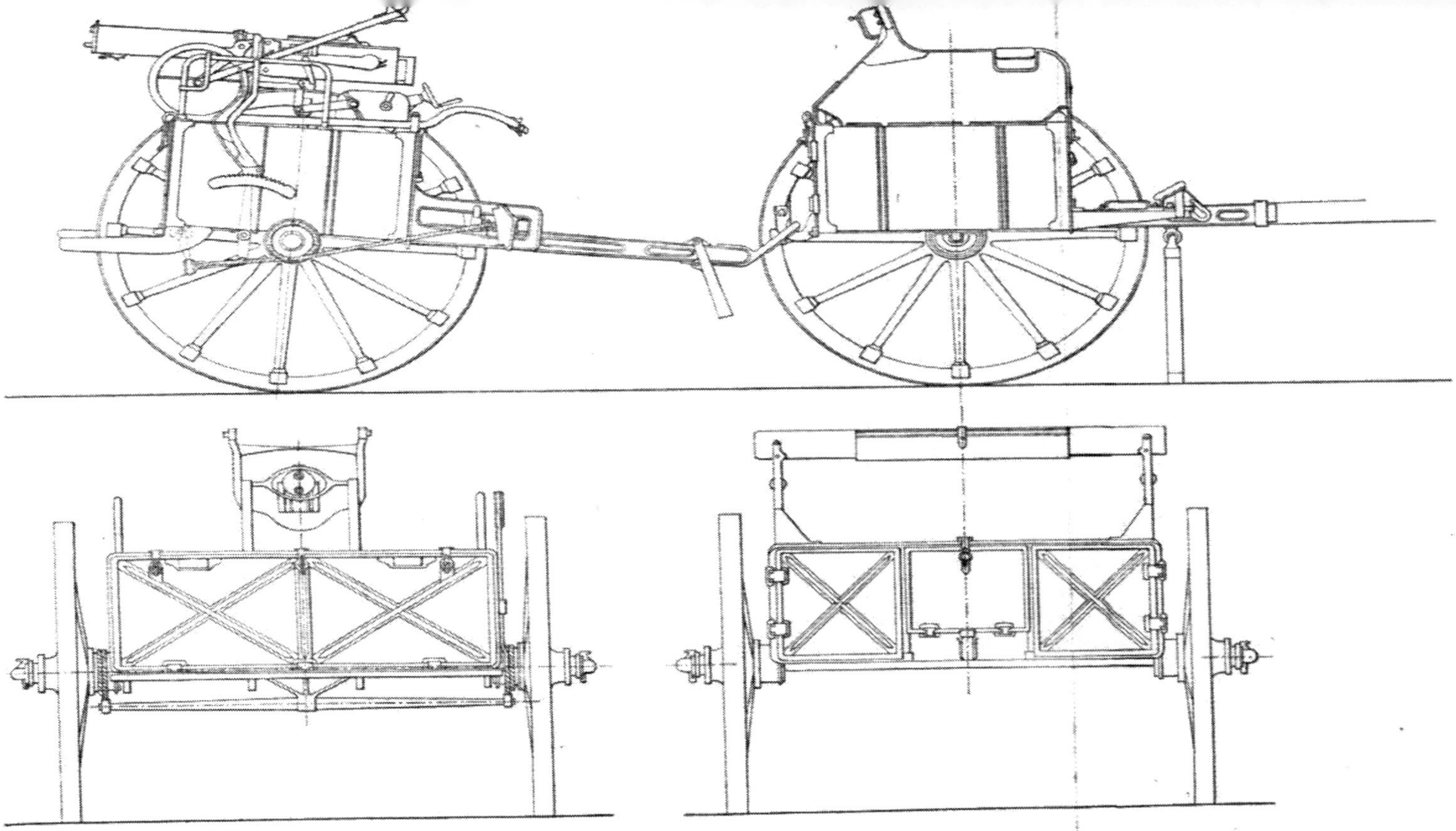

Limber and Carriage (to carry Five Gunners) for German Maxim R.C.A.M.Gun on Sledge Carriage.

Side elevation and rear elevation of carriage and limber. The belt-box sledge is secured in carriage under the gun.

Sketch from A. Korzen's work of 1908.

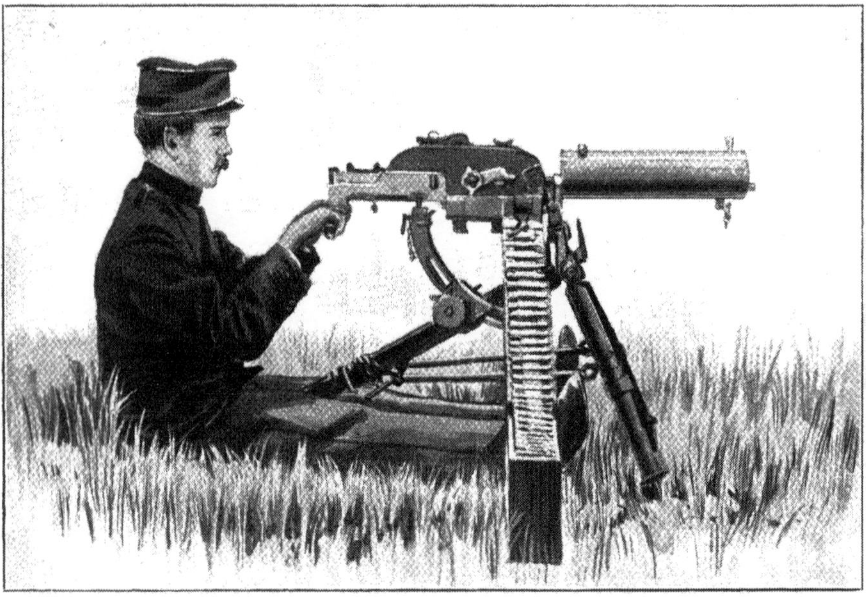

Schwarzlose R.C.A.M.Gun, which superseded the Skoda and Maxim Guns in the Austro-Hungarian Army during 1906.

Manufactured for the Government by a company at Steyer. Upper sketch shows tripod in lowest position, lower sketch shows the gun at usual height.

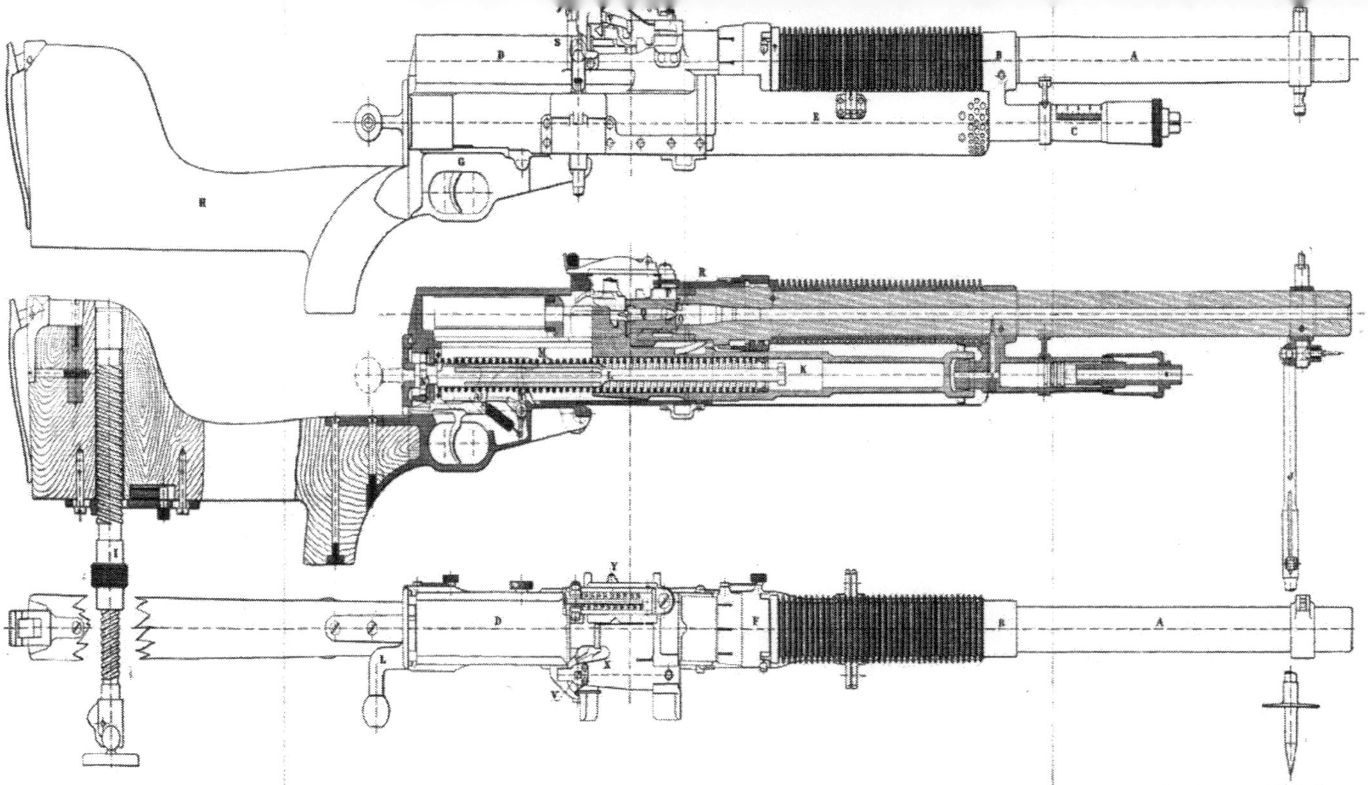

From a drawing by the French Hotchkiss Company.

Longitudinal Elevation, Section, and Plan of Hotchkiss Automatic Rifle, patented in 1907.

Came into the French service in 1908. Was tried in England during 1913. Now used by Cavalry. Worked by a powder-gas piston parallel with axis of the barrel. Fitted with open sights to 2,000 metres. Supported at muzzle and butt mechanically to assist the firer. The cartridges are held in metal feed-strips.

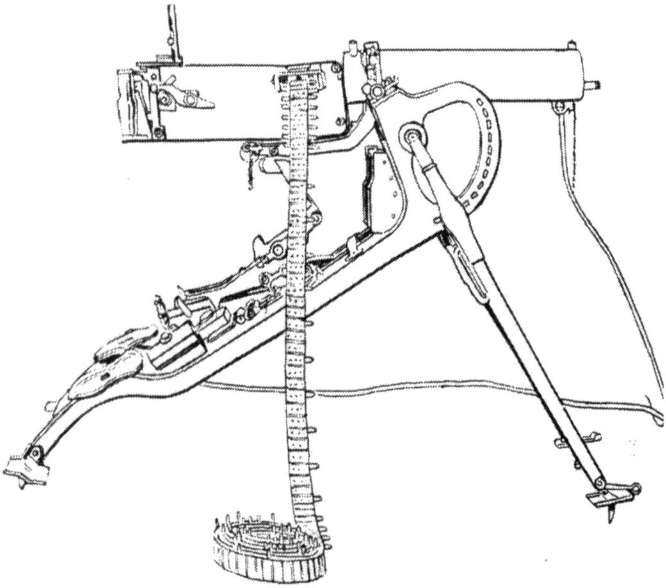

Extra Light German Sledge and Maxim R.C.A.M.Gun.

This gun was introduced in the German army in 1908. It weighs 24 grammes as against 56 in the preceding pattern.

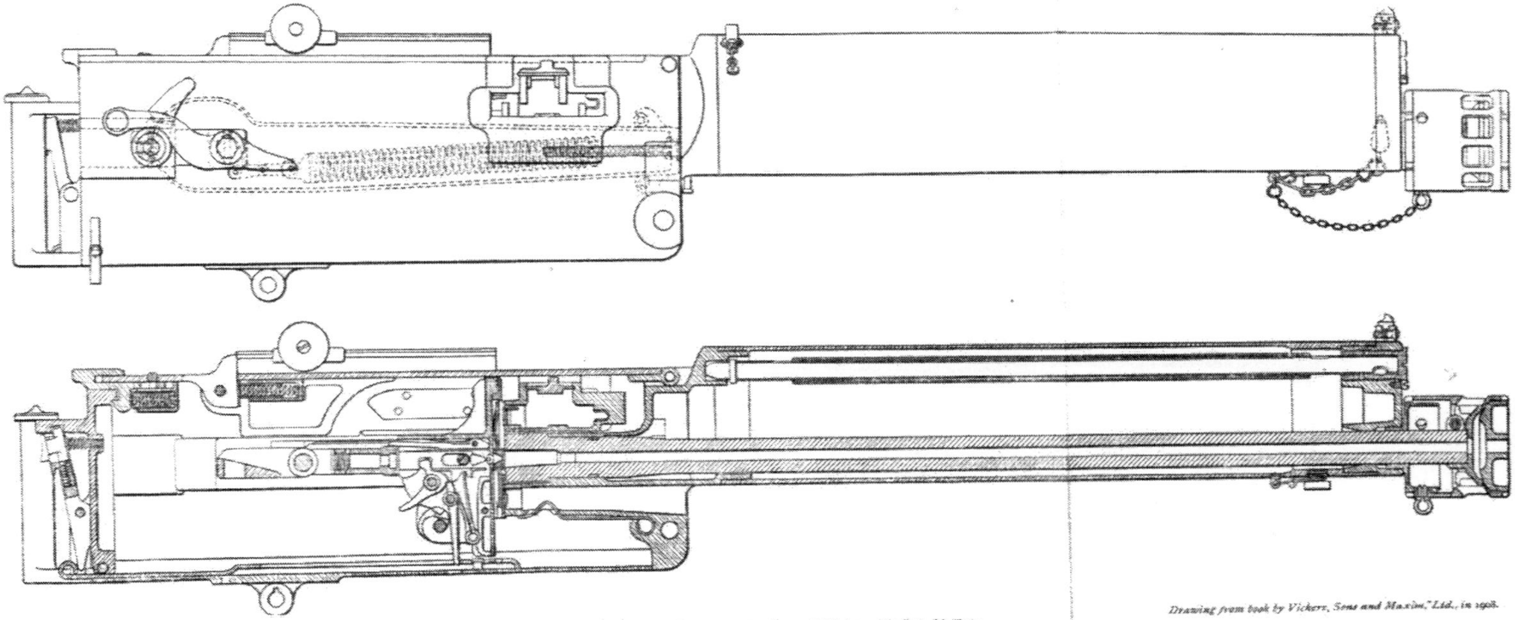

Longitudinal Elevation and Section of Vickers R.C.A.M.Gun.

Note steel water-jacket, simpler form of crank-handle, and no buffer spring.

Drawing from book by Vickers, Sons and Maxim," Ltd., in 1908.

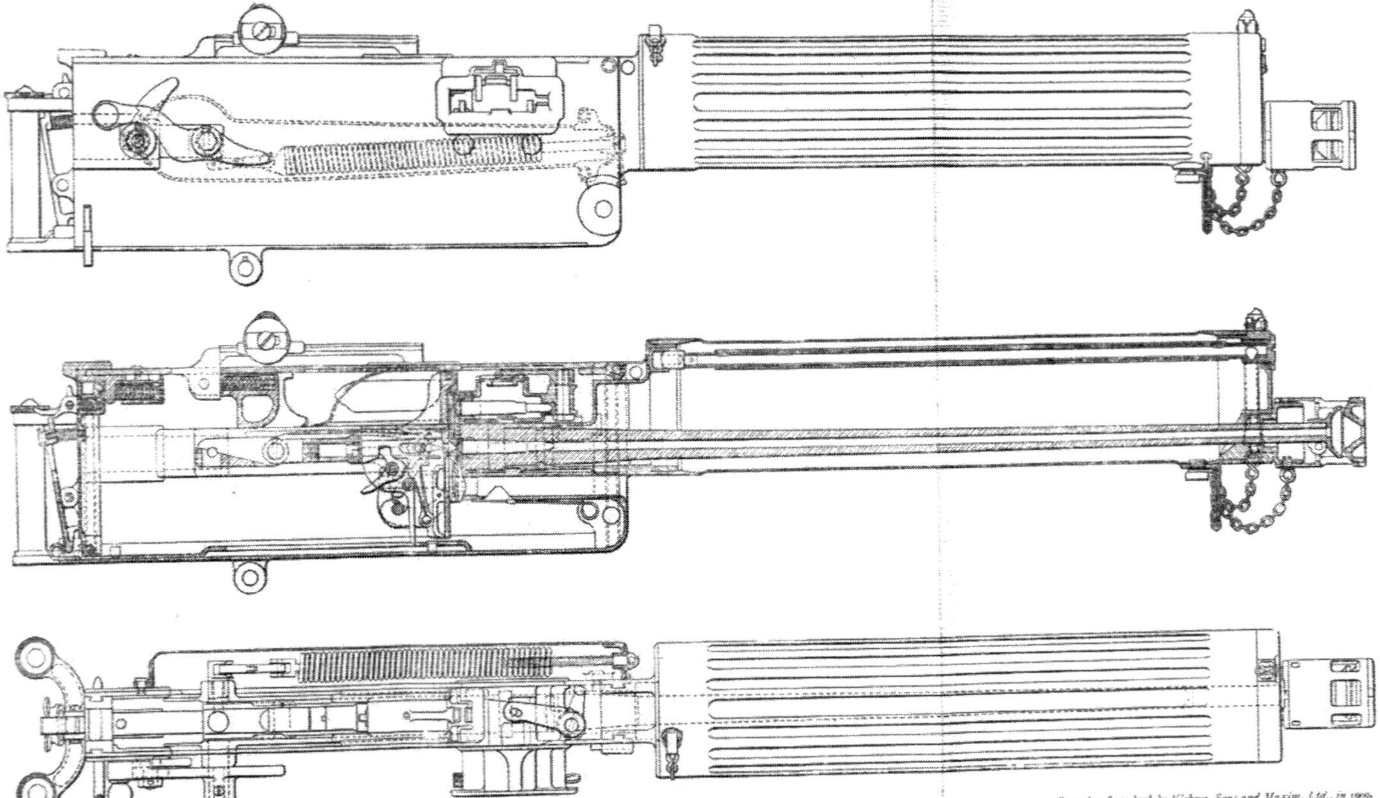

Drawing from book by Vickers, Sons and Maxim, Ltd., in 1909.

Longitudinal Elevation, Section, and Top Plan, of Vickers R.C.A.M.Gun with Corrugated Steel Water-Jacket.

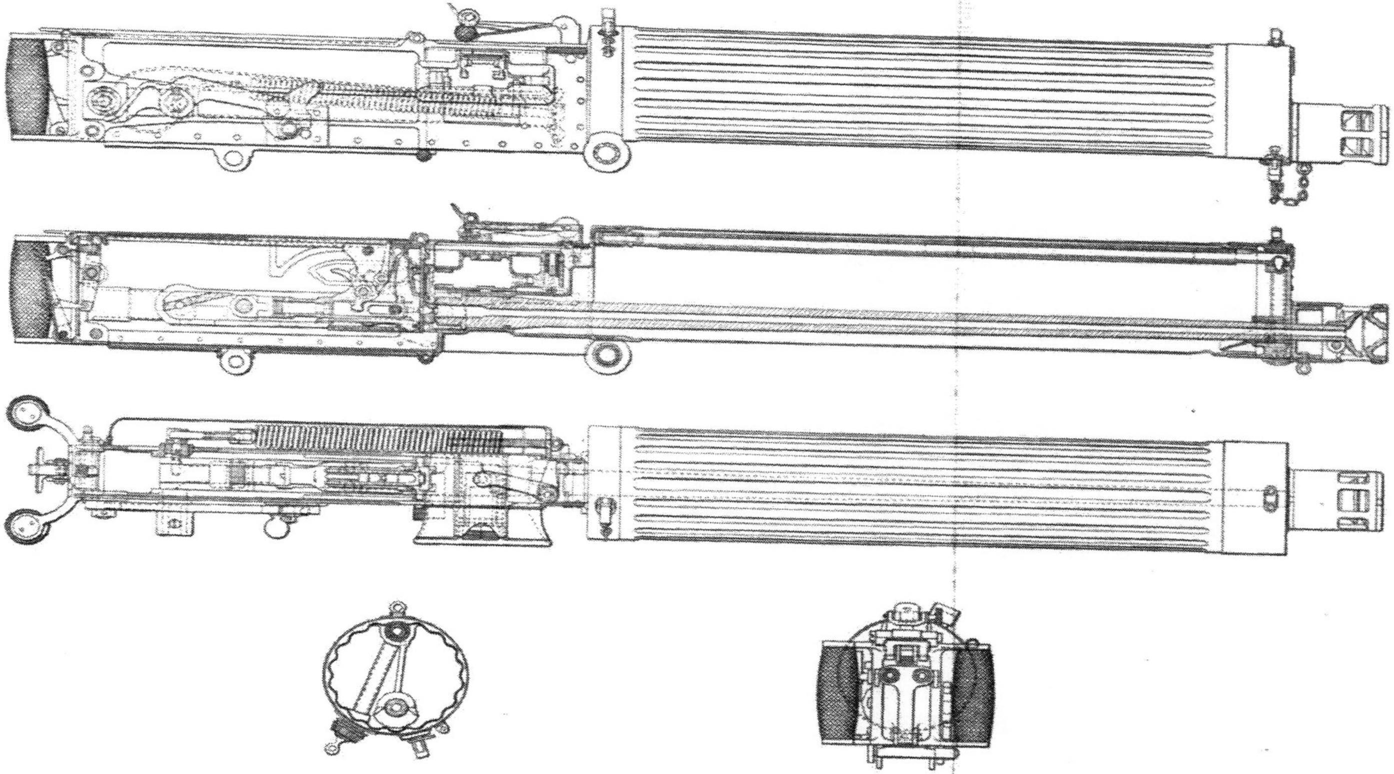

Drawing from book by Vickers, Ltd., issued in 1910.

Vickers (Light Model) R.C.A.M.Gun, showing Longitudinal Elevation, Section, and Top Plan, also Rear Elevation and Cross-Section through Water-Jacket.

Note lock is reversed, trigger bar is along the top of breech-casing cover, and the cover is hinged in rear of feed-block instead of in front of same; the back-sight is similar to that of our short rifle. A tangent sight was introduced by Vickers, Ltd., about 1913.

From photograph by Vickers, Ltd.

Tripod for Machine Gun mounted on Limber, in Accordance with British Patent Specification No. 11,253 of 1911. Manufactured by Vickers, Ltd.

From a photograph by Vickers, Ltd.

Tripod for Machine Gun mounted on Limber, in Accordance with British Patent Specification No. 11,253 of 1911. Manufactured by Vickers, Ltd.

The gun is being fired off the limber. Note the large number of belt-boxes, the smallness of the wheels.

From a photograph by Vickers, Ltd.

Tripod for Machine Gun mounted on Limber, in Accordance with British Patent Specification No. 11,253 of 1911.

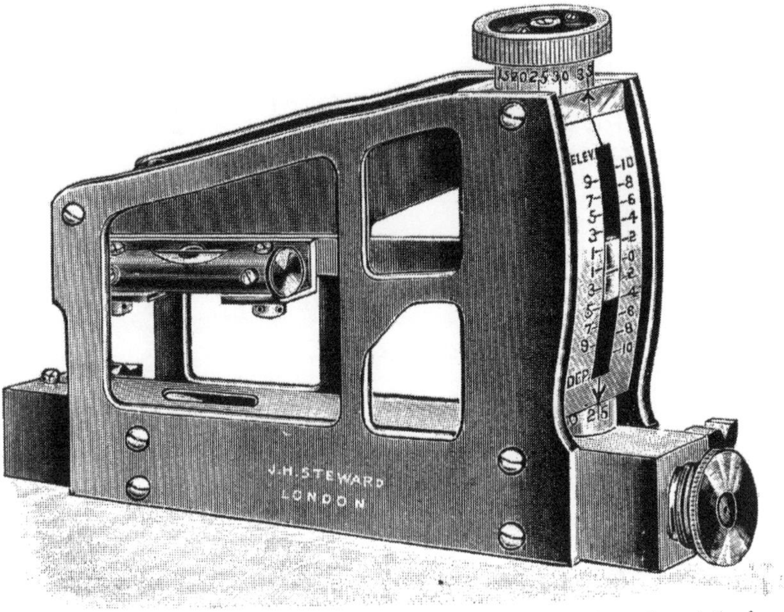

Block lent by Mr. J. H. Steward, 406, Strand, London.

The Combined Clinometer and Angle of Sight Instrument.

This was designed *circa* 1912, to supply the want of a small light instrument for use on the upper side of machine guns. The scale is divided into 10 degrees above (red) and the same below (black) the horizontal. Minutes are read from the graduated collars, above and below, every five only being given. Length of instrument is 5½ inches, weight 16 ounces.

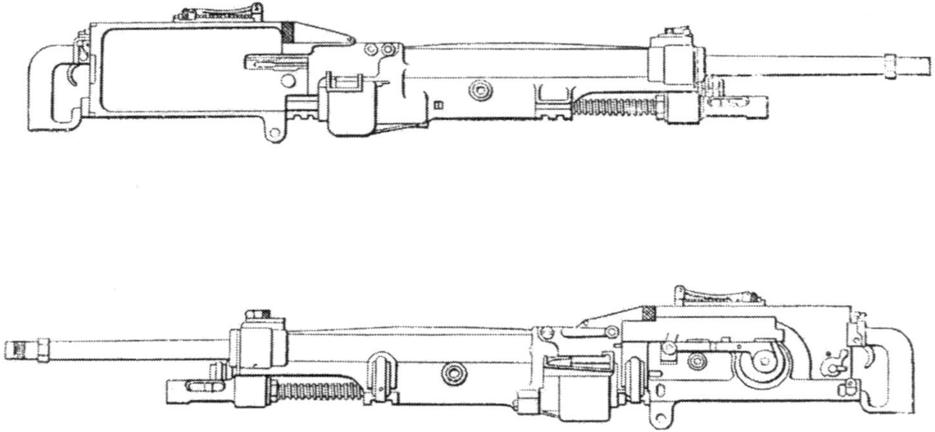

French Infantry Model 1907, R.C.A.M.Gun (St. Etienne), showing Longitudinal Elevations of Both Sides.

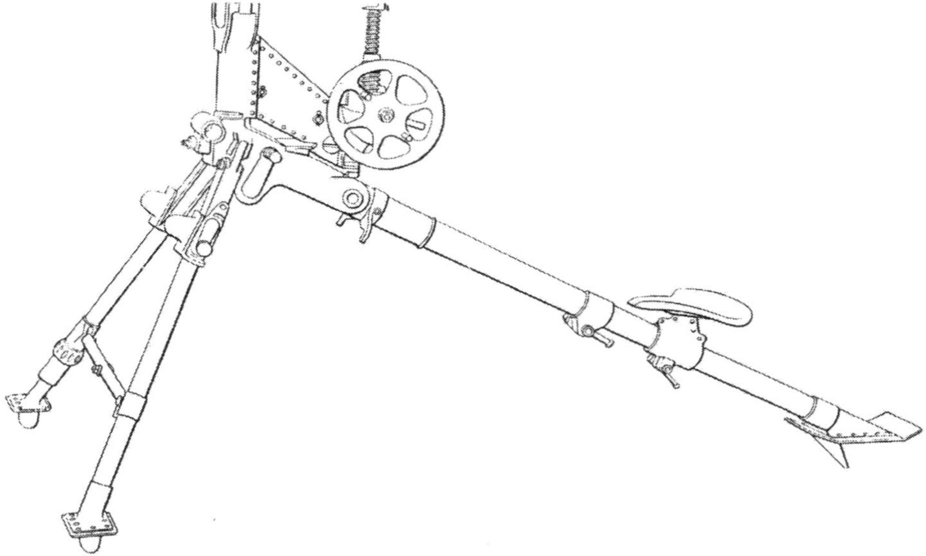

Tripod for French Infantry Model 1907, R.C.A.M.Gun (St. Etienne).

Note third leg is telescopic, the two front legs each have a knee-joint a few inches below the head, and that they will lie close together for pack transport.

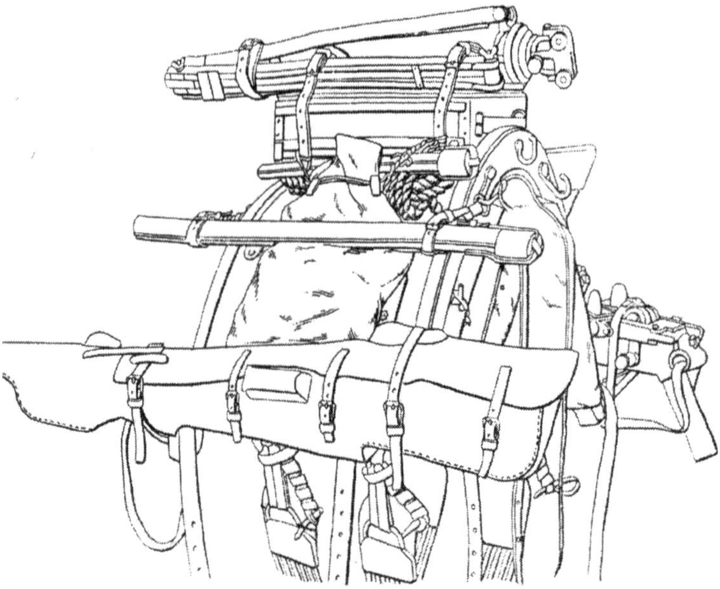

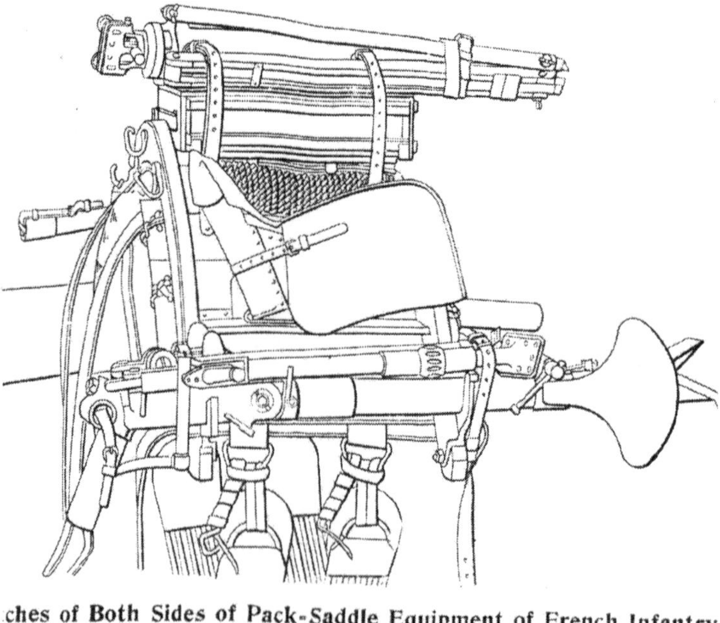

ches of Both Sides of Pack-Saddle Equipment of French Infantry Model 1907, R.C.A.M.Gun (St. Etienne).

Showing gun in case, tripod, head, spare barrel, range-finder, and tripod.

From a photograph by the French Hotchkiss Company.

The Improved Hotchkiss R.C.A.M.Gun of 1913, on its Tripod, with Empty Feed-Strips and Box.

The early pattern, of which this is an improvement, was brought out in 1896. Note shoulder rest, pistol grip, traversing gear and stops.

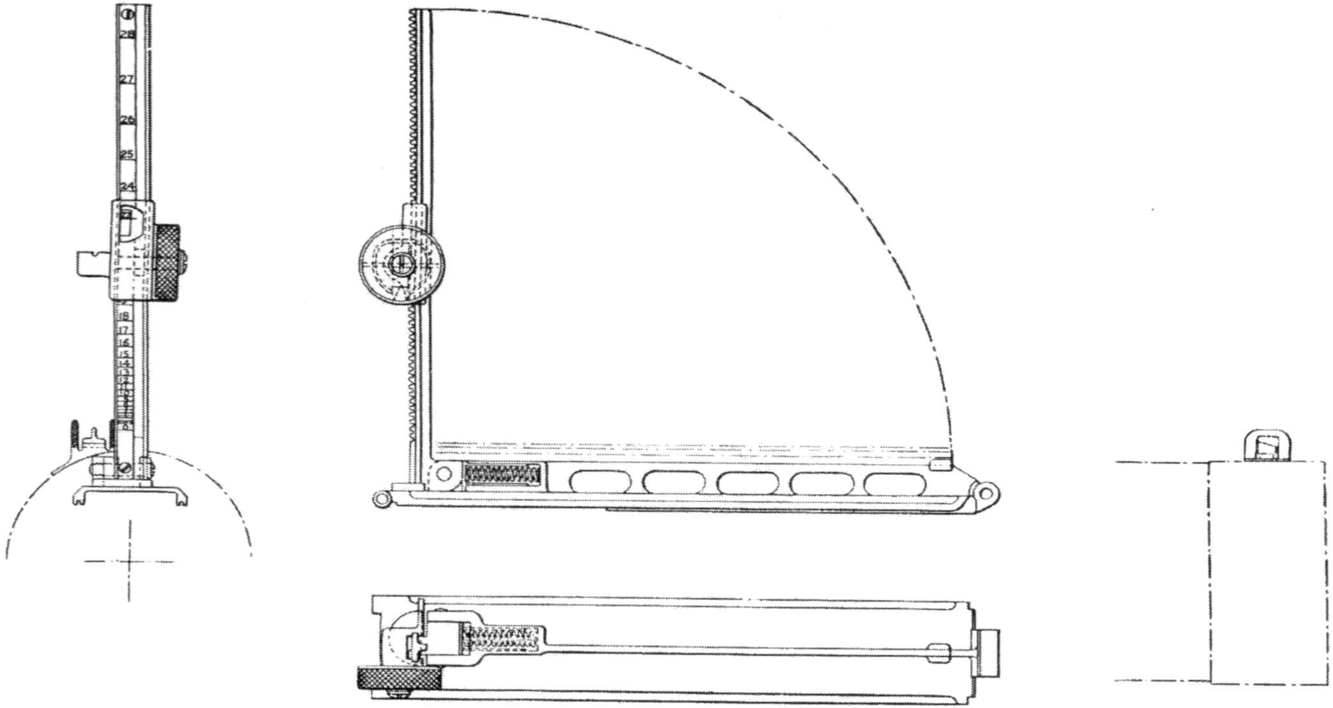

Drawing from book by Vickers, Ltd., issued in 1913.

Tangent Back-Sight on Breech Casing Cover, and Fore-Sight on Water-Jacket, the former being only graduated up to 2,800 Yards, while now (1916) the Canadians have used it up to 3,000 Yards.

This sight is for Vickers (Light Model) R.C.A.M.Gun.

From photograph by Vickers, Ltd.

Vickers Adjustable Tripod for Machine Gun (Vickers), to Specification No. 3,589 of 1913.

Tripod in lowest position. Note pipe from jacket to condenser, and tangent back-sight.

From photograph by Vickers, Ltd.

Adjustable Tripod, to British Patent Specification No. 3,587 of 1913 (Vickers).

From photograph by Vickers, Ltd.

Adjustable Tripod, to British Patent Specification No. 3,587 of 1913 (Vickers).

Tripod adjusted for use in a trench, with extension of third leg in use. This tripod has a traversing slide with two adjustable stops.

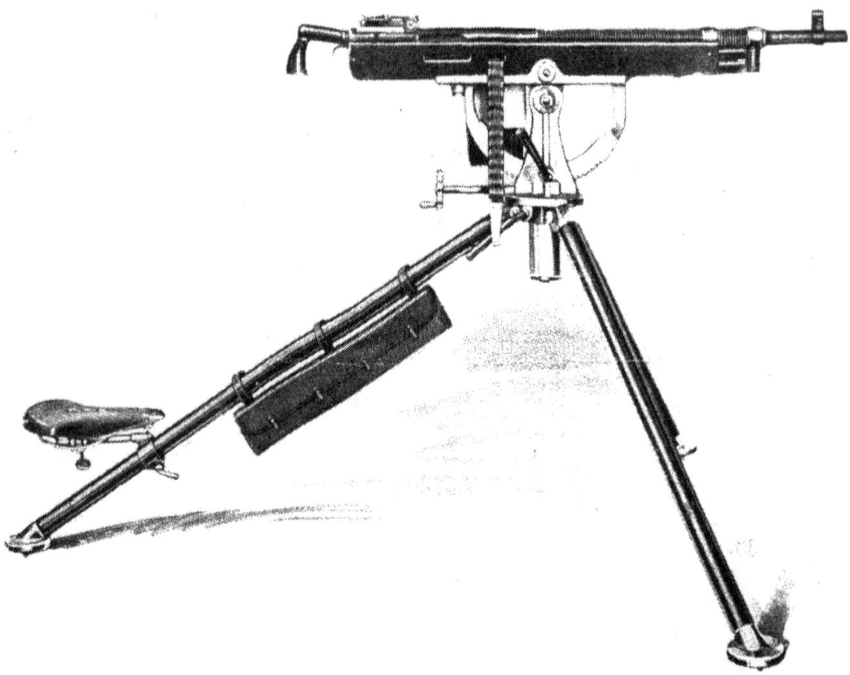

Photograph by Colt's Patent Firearms Manufacturing Co.

The Colt Rifle-Calibre Automatic Machine Gun, Model 1914.

The tripod is also the design of the Colt's Patent Firearms Manufacturing Co., Hartford, Conn., U.S.A. Note the legs of tripod are not adjustable for height.

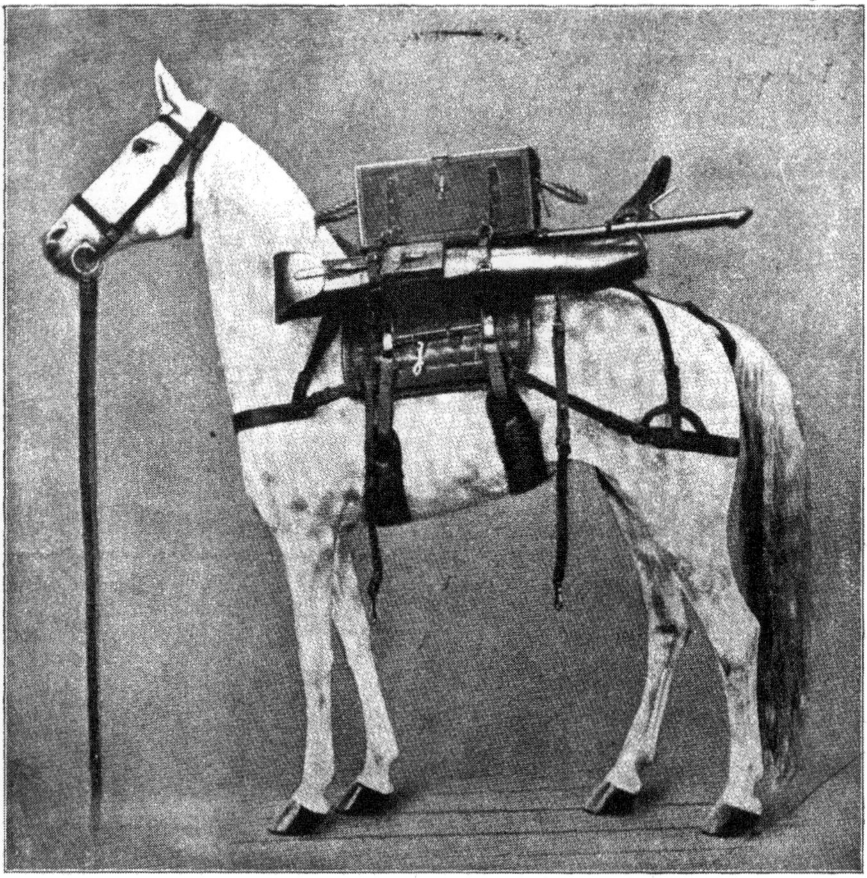

Photograph by Colt's Patent Firearms Manufacturing Co.

The Colt Automatic Machine Gun, Model 1914.

Pack-saddle equipment, gun in leather case, belts in box, and tripod on off-side.

Photograph by Gale and Polden, Ltd., Aldershot.

Cavalry Section at Firing Drill, Aldershot, with Vickers R.C.A.M.Guns, on Mark IV. Tripod, previous to August, 1914.

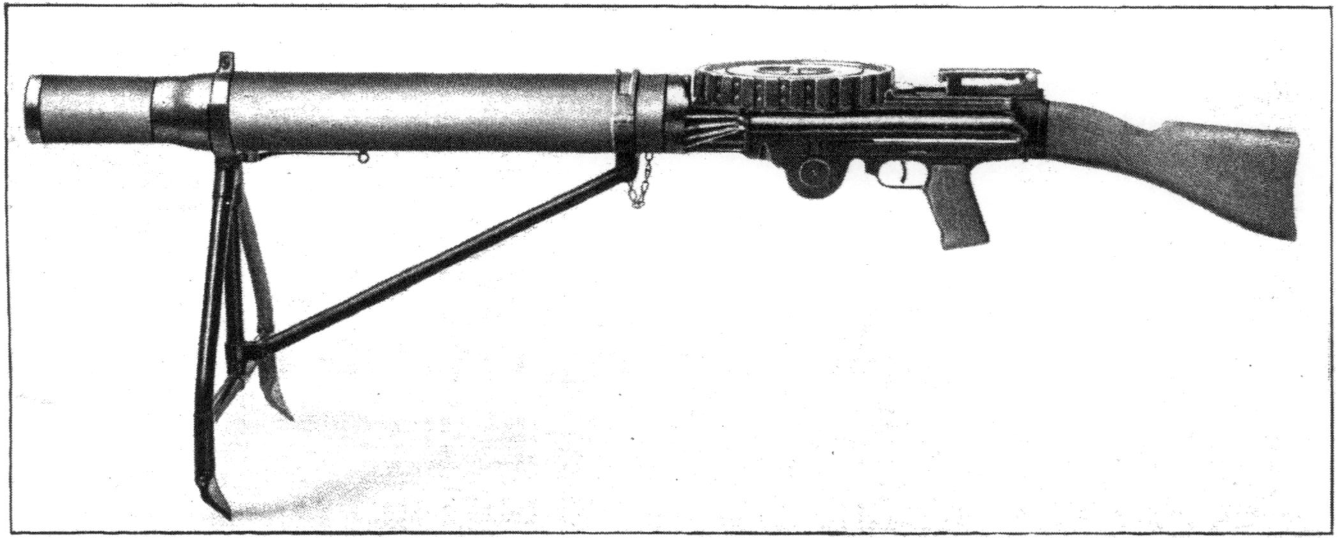

Photograph from Trade Handbook of the Lewis Gun, 1915.

The Lewis R.C.A.M.Gun, on Light Folding Field Mount.

There are also an aeroplane mount and a tripod mount, but the latter has not been made a Service issue. When a tripod is required, the Mark IV. is used with an adaptor.

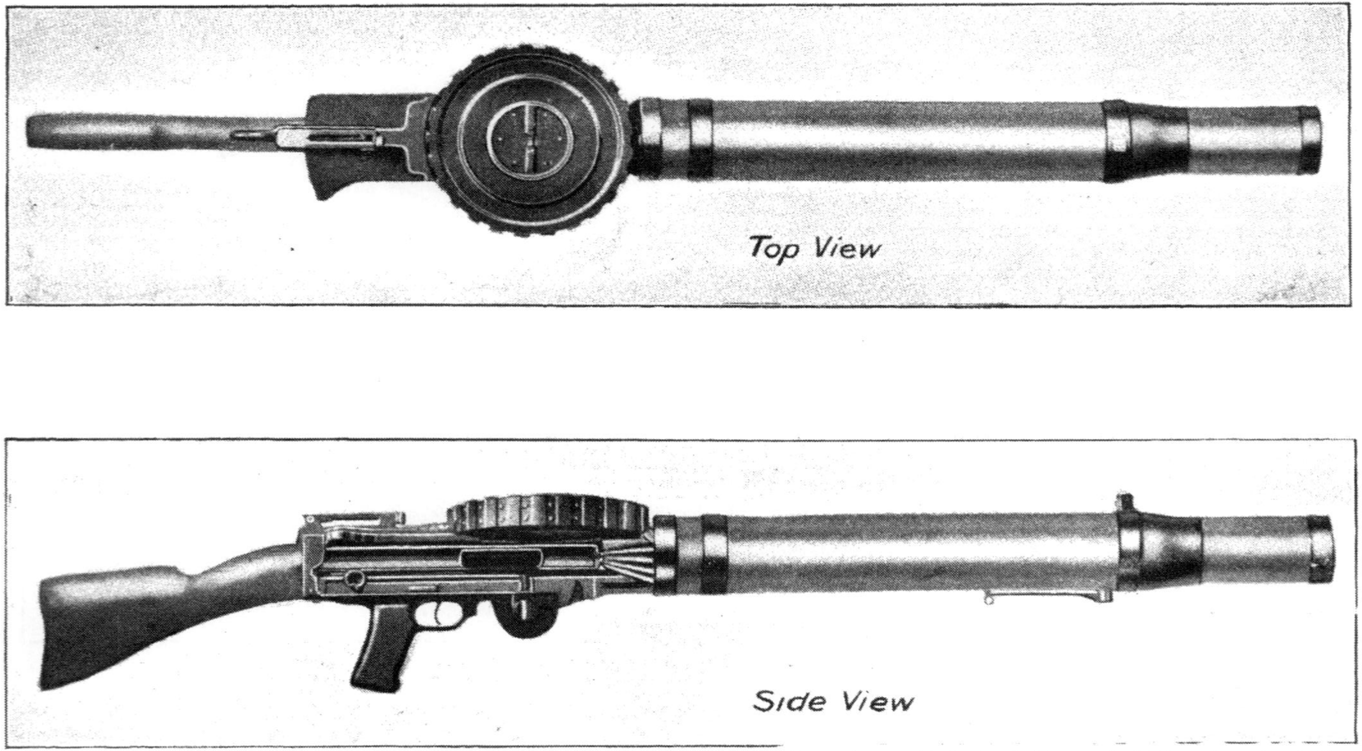

From photographs in Trade Handbook, Edition No. 4, Lewis Gun.

Top and Side Longitudinal Views of the Lewis R.C.A.M.Gun, English, ·303-Inch, Model 1915.

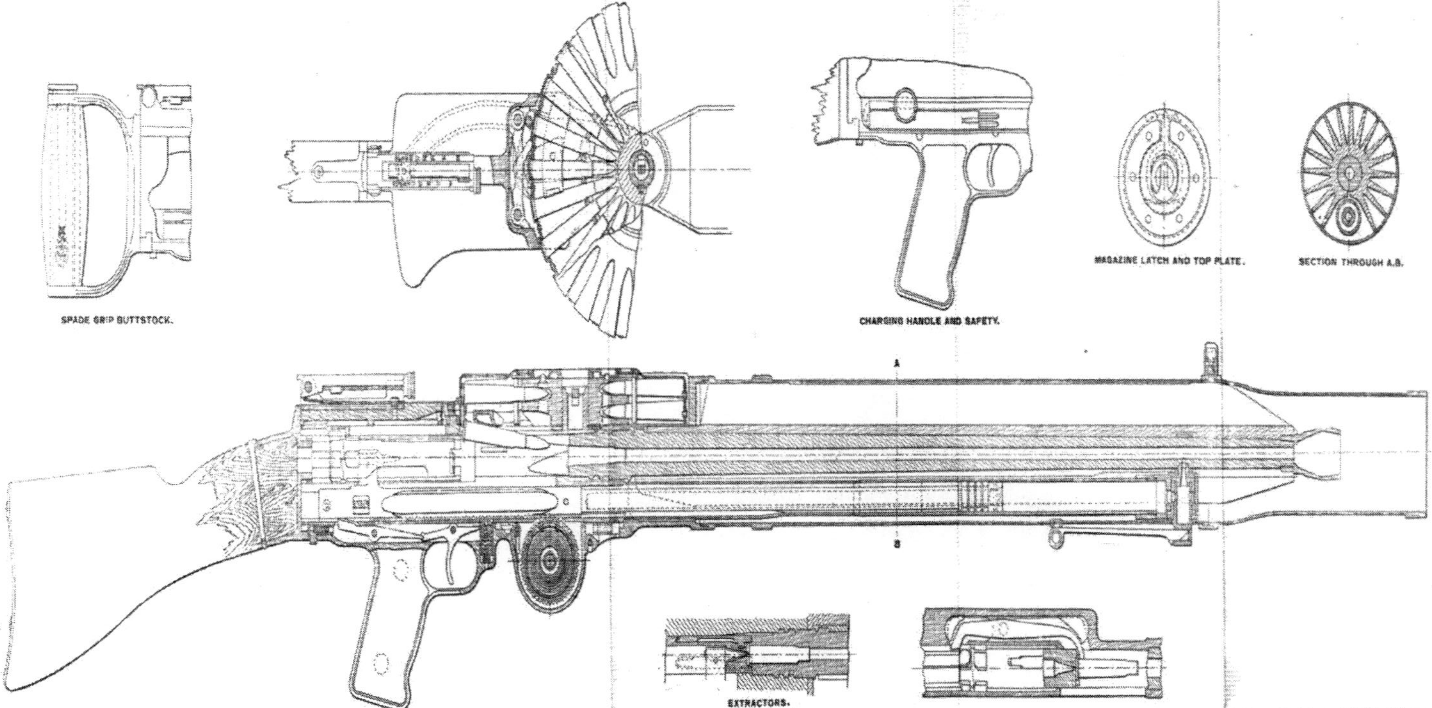

From drawing in Trade Handbook, Edition No. 4, Lewis Gun.

Partial Longitudinal Section of the Lewis R.C.A.M.Gun, English, ·303-Inch, Model 1915.

The rifle butt-stock and the spade-grip butt-stock are interchangeable. The two lowest sections give an empty case, in the chamber and extracted clear of same. The section through *AB* demonstrates the air-cooling system. The partial plan of magazine shows arrangement of cartridges.

From a photograph lent by "The Sphere.

The Maxim 1·457-Inch A.M.Gun (see illustrations of 1898), used in the Belgian Army (1915) against Aircraft from a Converted Field Artillery Trail Mounting.

Photograph from Record Press, 29, Fetter Lane, London.

Schwarzlose System R.C.A.M.Gun, captured from the Austrians; used by the Russians against Aircraft, about 1915.

Photograph from "The Sphere" of February 12, 1916.

The "Sokolov" Wheeled Carriage, Russian Army, with Vickers R.C.A.M. Gun and Shield.

[illegible] device with two wheels, a trail and two folding legs, and for the march is carried in a cart.

Photograph by Gibbs, Pirbright.

A Battery of the Motor Machine Gun Service (October, 1915) at Bisley.

The six guns are the "Vickers Light," each of which has a Mark IV. tripod in addition to the side-car mounting.

Photograph copyright, Underwood and Underwood London.

Maxim R.C.A.M.Gun in Turret on Armoured Car of British Machine Gun Corps (Heavy).

The officer standing on the car is General Smuts. Taken in East Africa by Lieut.-Colonel F. S. Keen, D.S.O. It appeared in *The Daily Graphic* of April 10, 1916.

No. 32

·303 VICKERS MACHINE GUN

1 Casing, breech
2 Casing, barrel
3 Bracket, foresight
4 Sight, tangent
5 Cover, front
6 Cover, rear (Lock rear cover)
7 Gland
8 Tube, steam
9 Grips, rear-crosspiece
10 Rear-crosspiece
11 *Arms of rear-crosspiece*
12 Lever, firing
13 Pawl, firing lever
14 Pin, screwed, axis, firing lever
15 Catch, safety
16 Pin, screwed, axis, safety catch
17 Pin, screwed, joint, rear-crosspiece
18 Slides, right and left
19 Roller
20 Handle, crank
21 Pin, screwed, fixing, crank handle
22 Bracket, elevating joint
23 Stop, mounting
24 Plate, bottom, breech casing
25 Bracket, check lever
26 Lever, check
27 Base of tangent sight stem
28 Bridge, rear cover
29 Pin, screwed, joint cover
30 Block, feed
31 Catch, front cover
32 Plug, screwed
33, 34 Thumbpiece, firing lever
35 Finger grips, safety catch
36 Pin-T, fixing, rear-crosspiece
37 Fusee
38 Chain, fusee
39 Box, fusee spring
40 Spring, fusee
41 Hook, fusee spring
42 Screw, adjusting, fusee spring
43 Lever of catch, front cover
44 Crank
45 Rod, connecting
46 Socket of side levers for 53
47 Tail of crank handle
48 Knob of crank handle
49 Casing, lock
50 Protector, screwed, condenser boss
51 Plug, cork
52 Plate, side, right
53 Stem of connecting rod
54 Tumbler
55 Protection on firing pin for 57
56 Trigger
57 Spring, lock
58 Sear
59 Extractor
60 Spring, gib
61 Trunnion block, barrel
62 Cannelure in 63 for asbestos packing
63 Barrel
64 Pin, axis, tumbler
65 Lugs on side levers for 83
66 Pin, split, keeper, bush, axis, side levers
67 Lever, top, feed block
68 Lever, bottom, feed block
69 Pins, split, fixing, top and bottom levers, feed block
70 Stud of top lever for feed block slide
71 Pawl, top, feed block, rear
72 Thumb grips of 71 and 73
73 Pawl, top, feed block, front
74 Spring, top pawls, feed block
75 Pawls, bottom, feed block (pair)
76 Pin, axis, bottom pawl, feed block
77 Finger plate of bottom pawls, feed block
78 Spring, bottom pawls, feed block
79 Levers, side (pair)

BREECH CASING PARTS. COVER REMOVED. BARREL CASING.

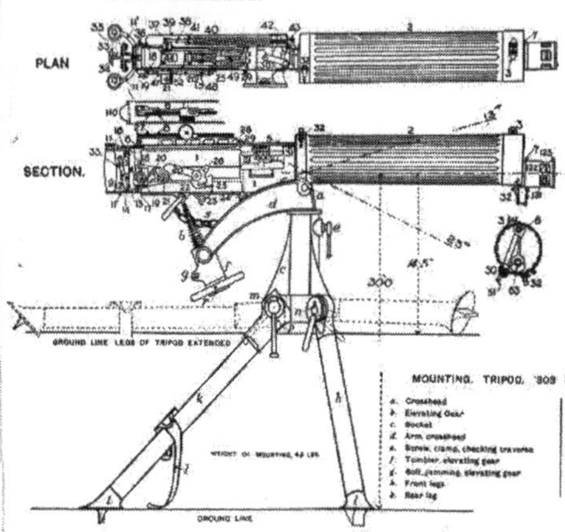

WORKING POSITIONS OF LOCK.

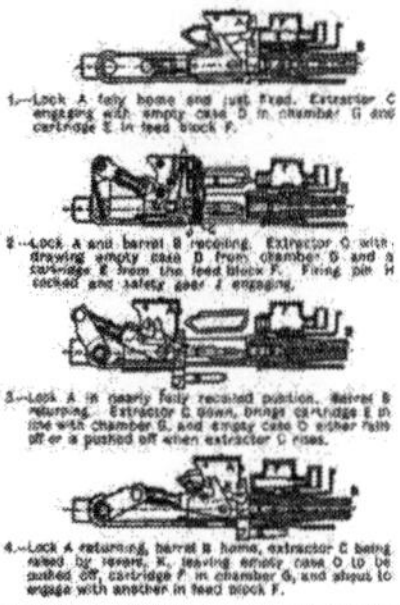

1.—Lock A fully home and just fired. Extractor C engaging with empty case D in chamber G and cartridge E in feed block F.

2.—Lock A and barrel B recoiling. Extractor C withdrawing empty case D from chamber G and a cartridge E from the feed block F. Firing pin H cocked and safety gear J engaging.

3.—Lock A in nearly fully recoiled position. Barrel B returning. Extractor C down, brings cartridge E in line with chamber G, and empty case D either falls off or is pushed off when extractor C rises.

4.—Lock A returning, barrel B home, extractor C being raised by levers, K, leaving empty case D to be pushed off, cartridge F in chamber G, and about to engage with another in feed block F.

MOUNTING, TRIPOD, ·303 INCH MAXIM GUN, MARK IV.

a. Crosshead
b. Elevating Gear
c. Socket
d. Arm, crosshead
e. Screw, clamp, checking traverse
f. Tumbler, elevating gear
g. Bolt, jamming, elevating gear
h. Front legs
k. Rear leg
l. Shoes
m. Socket legs
n. Stud, joints, and jamming handle, front legs
p. Joint pin and jamming handle, rear leg
r. Handwheel elevating gear
s. Chain, securing elevating screw
t. Strap to secure legs during transport

FEED BLOCK.

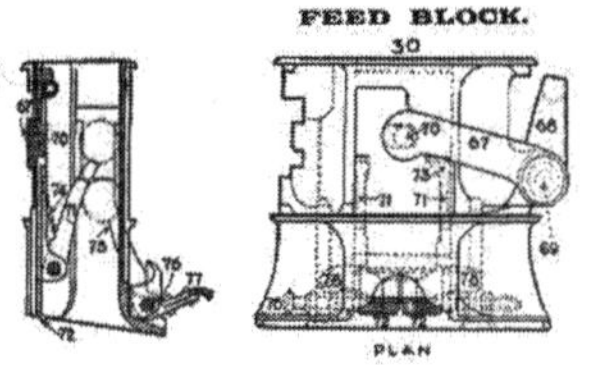

RECOILING PORTION OF GUN.

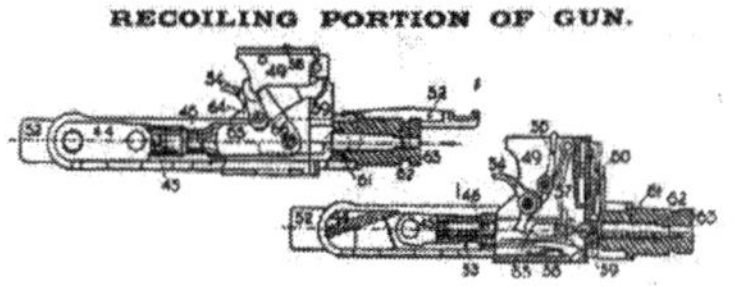

VARIOUS PARTS OF THE GUN.

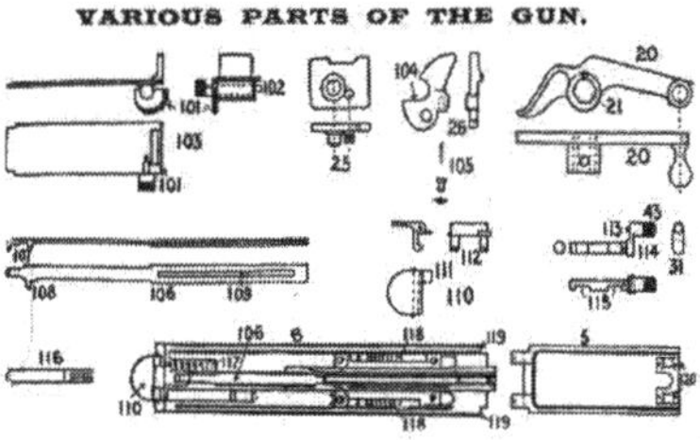

PARTS OF THE LOCK.

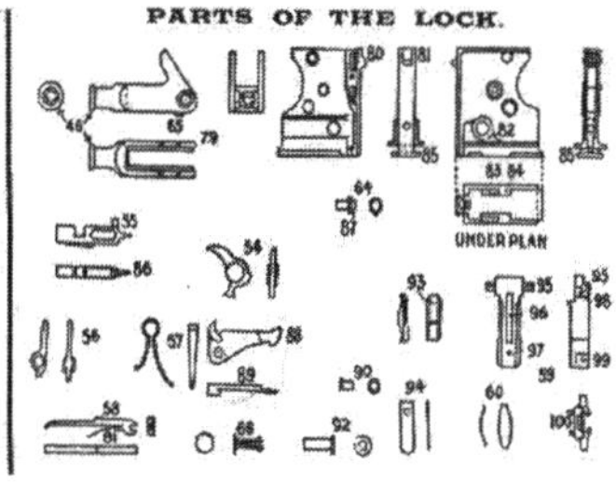

80 Upper extractor stop of lock casing
81 Lock
82 Bearings on lock casing for 79
83 Slots in lock casing for 65
84 Interruptions in flanges of lock casing
85 Flanges of lock casing
86 Pin, firing
87 Key of pin, axis, tumbler
88 Bent of extractor lever for 79
89 Lever, extractor, right
90 Pin, axis, trigger
91 Sear, spring
92 Bush, axis, side levers
93 Gib
94 Cover, gib spring
95 Horns of extractor
96 Recess in extractor for 93
97 Hole in extractor for 86
98 Shoulders of extractor for 89
99 Grooves in extractor for side plate springs
100 Grooves in extractor for 81
101 Thumb piece, sliding shutter catch
102 Plunger, sliding shutter catch
103 Shutter, sliding
104 Pin, keeper, check lever
105 Piston, check lever / Spring, check lever
106 Bar, trigger
107 Lug on trigger bar
108 Lug on trigger bar for 117
109 Slot in trigger bar for 56
110 Lock, rear cover
111 Lug on rear cover lock for 116
112 Hooks of rear cover lock
113 Plunger, front cover catch
114 Hole for keeper pin, front cover catch
115 Grooves in front cover catch to clear 120
116 Spring, rear cover lock
117 Spring, trigger bar
118 Ramps, rear cover
119 Grooves in rear cover
120 Hoops of front cover catch
121 Cup, muzzle attachment
122 Casing, outer, muzzle attachment
123 Cone, front, muzzle attachment

Printed and Sold by GALE & PO[illegible] Ltd., Wellington Works, Aldershot P5970.

www.ingramcontent.com/pod-product-compliance
Ingram Content Group UK Ltd.
Pitfield, Milton Keynes, MK11 3LW, UK
UKHW042100021026
15589UKWH00001B/82

9 781843 425595